Old Farm Tools and Machinery

By the same author:

Country Craft Tools

Old Farm Tools and Machinery
An Illustrated History

by
Percy W. Blandford

GALE RESEARCH COMPANY
FORT LAUDERDALE, FLORIDA

Printed in Great Britain for
David & Charles Limited Publishers
Brunel House   Newton Abbot   Devon

Published in the United States
by Gale Research Company
Fort Lauderdale, Florida

Published in Canada
by Douglas David & Charles Limited
1875 Welch Street   North Vancouver   BC

Library of Congress Cataloging in Publication Data

Blandford, Percy W
  Old farm tools and machinery.

  Bibliography: p.
  Includes index.
  1. Agricultural machinery--History. 2. Agricultural imple-
ments--History. 3. Agricultural machinery--Pictorial works.
4. Agricultural implements--Pictorial works. I. Title.
S674.5.B55   1976   631.3'09   75-44376
ISBN 0-8103-2019-3

Contents

List of Illustrations		7
1	The Rural Scene	11
2	From Animal to Steam Power	16
3	The Coming of Tractors	30
4	Ploughing	43
5	Planting and Sowing	74
6	Cultivating	86
7	Harvesting Cereals	113
8	Harvesting Other Crops	134
9	Estate Management	148
10	Stock Management and Feeding	155
11	Dairy Produce	166

Appendices
1 Sources of Information 175
2 Glossary 177

Bibliography 183

Acknowledgements 184

Index 185

List of Illustrations

PLATES

1 Horse-driven sorghum press — 17
2 Gloucestershire waggon — 22
3 Threshing box and steam engine — 25
4 'Titan' tractor — 37
5 'Farmall' three-wheel tractor — 38
6 Walking tractor cultivator — 39
7 Modern tractor with fertiliser attachment — 41
8 The action of ploughing — 46–7
9 The action of a plough — 48
10 Kent turnwrest plough — 52
11 Sussex foot plough — 54
12 Rutland plough — 58–9
13 Turn-over horse plough — 63
14 Two-furrow horse plough — 64
15 Ridging plough — 65
16 Modern multiple turn-over plough — 72
17 Wheels of a potato planter — 76
18 Mangold drill — 78
19 Jabez Buckingham root drill — 81
20 Post-hole borer — 97
21 Wooden harrow — 103
22 Spring-tine harrow — 103
23 Horse-drawn hoe — 111
24 Sail reaper — 121

7

25 Massey-Harris binder   126
26 Grain threshing in Turkey with oxen   127
27 Threshing machine   130
28 Forage harvester   141
29 Horse-drawn nib   153
30 Chaff-cutter used as museum sign   160

DRAWINGS
 1 Using horse power   18
 2 Carts and waggons   21
 3 Ploughing engine   27
 4 Four-stroke cycle   31
 5 Primitive ploughs   43
 6 Plough parts   45
 7 Wooden ploughs   51
 8 Guernsey plough   55
 9 Iron ploughs   57
10 Variations on the plough   62
11 Steam ploughing   69
12 Tractor ploughs   71
13 Planting tools   75
14 Planting devices   80
15 Fertilizer spreading   83
16 Spades   87
17 Forks and hoes   89
18 Cultivating tools and rakes   92
19 Hoes and billhooks   95
20 Rollers and levellers   99
21 Harrows   101
22 Disc harrows and cultivators   106
23 Diggers and hoes   109
24 Hand reaping tools   114
25 Reapers   118
26 Hand harvesting tools   123
27 Self-binding reaper   125
28 Threshing and winnowing   128

8

29 Combine harvester parts   131
30 Mowing and swath turning   135
31 Elevators, baler and forage harvester   139
32 Potato harvesting   143
33 Beet harvesting   146
34 Hedge cutting and fencing   149
35 Forestry tools   151
36 Stock handling tools   156
37 Food-preparing equipment   158
38 Barn machinery   163
39 Milking and veterinary tools   167
40 Dairy equipment   169
41 Cheese equipment   173

While the earth remaineth, seed time and harvest, and cold and heat, and summer and winter, and day and night shall not cease.

(*Genesis* VIII, 22)

CHAPTER 1

The Rural Scene

Man has always had to depend on the land for food. Except for fish from the sea, he could not survive without the products of the soil, acquired either directly through plant life or from animals that live off plant life. He is dependent on nature. He relies on the seasons to occur regularly and bring forth harvest. Farming might be regarded as the using of nature to guide her products more towards the needs of man; directing their course so as to channel them more conveniently to his needs.

Man is not always successful. He does not always get what he wants. When the monsoons fail in India, when there are floods in Bangladesh, or when he tries to take everything and put nothing back, as in the dust bowls of the central USA, he is brought up with a start, and he has to rethink. He has to be attuned to nature and work with her.

Earliest man was nomadic and took what he could find, gathering berries and fruits while hunting animals. Nature supplied and he took, and as long as the population was sparse and he was prepared to travel, he could live like that. There still are primitive people living in this way, but their numbers are rapidly diminishing.

Tools, as such, were unnecessary for these hunters and foragers. They had weapons to kill animals and sticks and branches on the spot might be used to reach fruit.

With the urge to settle down, and live with more stability in families or communities, must have come a realisation that instead of roaming in search of food it would be better to produce food on the spot, and this was the start of cultivation and the domestication of animals.

With his land to cultivate and herds to maintain, the first primitive farmer had to devise and make tools, some of which would have been adaptions of his weapons. He had to fence his land to keep his animals in

11

and wild animals out. He had to learn how to plant, to discover what crops yielded the best results, to learn to till the soil. He had a lot to accomplish, for he was pioneering what is the biggest industry in the world today—it has to be, otherwise man could not survive. Men who found they were particularly good at making things became specialists. They were the first of the country craftsmen and their skills founded the activities which have developed into the giant international agricultural manufacturing and engineering industries of today.

We have much to thank the blacksmith for. Ever since iron was first produced there have been craftsmen who used fire and their own skill to fashion it into implements. The smith's trade has changed little. The early smiths made their own tools as well as those for the farmer and other craftsmen. The tools of a modern smith could be recognised and utilised by a smith of Biblical times.

The other great craftsmen who helped the farmer were the woodworkers. They have seen greater changes. Where they once had to rely on crudely hacking wood, they have progressed, through more precise hand tools in the Middle Ages, to mechanical tools that reduce labour and do the routine jobs more accurately, although there is still opportunity for the exercise of much skill.

The history of farm implements from the earliest days to the present makes a fascinating study. There was a time when the majority of men and women were involved in agriculture. Numbers have diminished—very rapidly after the first quarter of the twentieth century—yet a very small number of workers are now producing a considerably greater amount of food from the land. The reason is machinery. Although comparatively few workers are needed to use the machinery, they have to be backed by large numbers who build and maintain it. What they have made over the centuries, and what they are making now, add up to the subject of this book.

Progress in farm mechanization has been greater over the last century than in the whole history of farming before that, because of the improved power sources available. When the only source of power was horse or ox there was a limit to the machines that could be devised. While a horse is stronger than a man, and several could be used together, the resulting power could not compare with that from steam or oil engines—which do not tire. During the horse era some large and clever agricultural implements were devised, but none of them had widespread use. Agricultural work continued to rely on many men and many horses.

While Britain continued a way of life that had seen few upheavals for thousands of years, much of America was in a pioneer state. This had two effects. In the early stages pioneer farmers had to make and devise farm implements from what was to hand, while British farmers were beginning to benefit from the mechanization of the Industrial Revolution. This meant that some more basic implements were still in use in America perhaps a century after they had been superseded in Britain. By the time farming had become established in the furthest parts of the USA, industry had also been set up and the country was no longer dependent on Britain for machines.

The result was a move from Britain to America for the development of larger and more complex machinery. In that vast country the areas to be worked were much bigger and the potential for big and expensive equipment was greater. Many of the machines were too big to be justified for the smaller areas to be dealt with in Britain. It was not until the middle of the twentieth century that more modest-sized machines for the same purposes began to be used in Britain.

Today, the operator in his air-conditioned cab may be doing the work of ten men in a tenth of the time, while making use of computerised data and scientifically formulated seeds and fertilizers. He may be highly trained and his technical ability may be such that he would amaze a man on the land only a decade before, while the farm worker of a century ago could not begin to comprehend what was going on. But what is going on?

It has to come back to nature and the seasons. However mechanised the farmer becomes, he has to go along with nature. The singer who wants to 'Plough and sow, and reap and mow, and be a farmer's boy' sums it up. The sequence and the seasons must coincide.

The land has to be prepared. Although there have been attempts to prepare land for a crop without disturbing it, the accepted way is to open it and leave it exposed. No-one has found a better way of doing this than with a plough—one of the earliest implements. Although there are large numbers of ploughs described later, the tools used today are recognisably the same as those of hundreds of years before. Ploughs have not been developed out of all recognition, or superseded, as have other processes.

Planting and sowing have changed dramatically. From a hand process with simple equipment that was wasteful and slow, seed is now planted with a mechanical precision that gives a spacing and depth previously calculated and may also deposit a measured amount of water or specially compounded fertilizer with it. Coupled with this is a reduction of labour and a great saving in time. The end product is a better crop.

It is probably in the gathering of the harvest—whatever the crop—that there have been the greatest changes. Early farmers were very much at the mercy of the climate, and getting in many crops was a race while the weather was favourable. The harvest thanksgiving service in church was a very real thanksgiving, and the harvest supper provided by the farmer for his men was a real sign of gratitude and relief. No doubt today a farmer is thankful and relieved when the crop is in, but his risk of loss or failure is very slight even compared with the days of his father.

The cereal crops were harvested painfully slowly by hand cutting, gathering into stooks to dry, then carting to be stacked in a barn and threshed later. Threshing had the advantage of providing employment in the winter when there was little else for a man to do, but regarded as a production exercise the whole process was very poor. The modern combine harvester eliminates several steps as it eats its way into the standing crop and delivers the grain without stopping, doing the work of dozens of men and women at a tremendously faster rate.

With animals in general, mechanization has not been so pronounced. What developments there have been have tended towards factory methods rather than the use of implements. Sheep, and many other animals, still roam the fields, but with increased scientific knowledge the move is towards intensive feeding and rearing.

The milking of cows has changed almost completely, from hand methods with simple equipment, to complex machines, and the further factory stages of converting milk to butter and cheese are equally complex. Hens no longer scratch in the farmyard and provide pin money for the farmer's wife with their eggs. Instead they have become almost egg-laying machines in cages, with food entering at one end and eggs emerging from the other. Calves, pigs and other animals reared for meat may never see a field or even natural light. Whatever the rights or wrongs of this, it is a far cry from the farming methods of only a few years ago.

The rural scene too has changed. Whatever nostalgia may conjure up about the pastoral scene as it was, there is no going back. For most of its history farming has been largely a personal hand to land operation, with simple hand tools or rather basic implements and no more power than could be produced by the animals which were used.

Modern farming is a complex business, but the general problems of the farmer are not the concern of this book. Some present-day methods may be controversial. For farming methods, choice of crops, opinions on techniques and anything other than the subjects embraced by the title, the reader must search elsewhere. But for anyone interested in all kinds of tools and implements used in the production of food from the soil, this book covers developments from the earliest days and most primitive methods, through the gradual acquisition of mechanical knowledge and skill to present-day power farming.

CHAPTER 2

From Animal to Steam Power

In the earliest days of cultivation, when man started to wrest crops from the soil and domesticate animals on it, he relied mainly on his own muscle power to do the work. Examples of early implements show the application of the lever, and provision for numbers of workers to share in providing greater power than one man could exert. Early man soon put his hand-held flint on a handle to get the extra power that would come from a swinging action, and digging sticks were developed into such tools as the cas chrom (Fig. 5, p 43).

The first plough may have been the result of an attempt to use several workers to drag a digging stick through the soil to break it up. This would have been arranged with one man steering and many others pulling. Such a plough probably pre-dated the use of the wheel. Basic ploughlike tools are shown in early Egyptian drawings, and these would have been used before later civilisations used the wheel.

Man used the animals he had tamed such as oxen and bullocks to provide extra agricultural power. Oxen are still used in many countries, being comparatively placid and easily trained and managed for simple pulling work where they can exert considerable power, even if their speed is slow. They have a neck and shoulder formation which permits a simple transfer of their effort to pulling an implement via ropes or chains attached to an uncomplicated yoke, which is very much simpler than the harness used to transmit the pulling power of a horse.

Man probably used the horse first for riding and then pulling carts and waggons on roads, before they found much use on the land. During the very long period in which horses have thrown in their lot with man, they have been bred to develop certain attributes. A horse intended to pull a load or otherwise provide power is a very different beast from a hunter or

16

Plate 1 Sorghum press in Florida, USA. A horse walking in a circle at one end of the
beam turns the vertical rollers which squeeze juice from the cane

racehorse. A horse intended for pulling a cart or other light vehicle is of
medium size, capable of exerting the required power and travelling at a
reasonable speed. For work on the land, or for pulling heavily loaded
farm waggons, the accent has to be on strength without much speed.
This has resulted in the development of such breeds as the Shire, Clydes-
dale, Suffolk and Percheron—all large powerful horses, with the first
two, at least, developed from the massive war horses needed to carry the
heavy plate armour of medieval days. The Percheron is of French origin
and did not find its way into Britain until being bred for military purposes during World War I.

The usual employment for a horse was in giving a straight pull. For a
heavy load, two or more horses could be used side by side or in line
(tandem). For some implements there were shafts similar to those used
on a cart, but for those such as ploughs the pull was via chains and whippletrees (Fig 6T, p 45). If large numbers of horses had to be used there
was always the problem of the damage they might do with their hooves,
and several men might be needed to control them.

For other power applications other ways had to be devised to use horse
power. A load could be lifted by a horse walking away, pulling a rope
over a pulley, and there were also methods where a horse might work
directly, but the snag with this could be the lack of precision and the
absence of control, so that a load could fall, a device be strained, or work
damaged. Another way of using the horse to produce power was to have

17

him walk in a circle, harnessed to an arm on a vertical axle.

A simple application came about when the vertical axle could be made to do the desired job. In the American sugar, or sorghum, press (Plate 1) the shaft formed a sort of vertical mangle through which the cane was passed to squeeze out juice, which ran off to a container. The arm to the horse was arched high enough to clear a man sitting to operate the press.

A more permanent assembly on some larger farms had the horse walking a circular track under a large toothed wheel (Fig 1A). This acted as a source of power for barn and dairy machinery. The wheel, mounted on a substantial central axle, had teeth engaging with one or more smaller wheels and these might be coupled to a further train of gears, either cast-iron or wooden-toothed. One model, now in the Science Museum, London, was in use at Broughton Manor Farm, Bierton (Buckinghamshire) until 1879. The large wheel, about 16ft in diameter, has 240 teeth, while succeeding gears have 40, 48 and 24, giving an overall ratio of 12:1. In the example, the wheel is coupled to a 60-gallon butter churn. The horse, walking at 2mph, turned the big wheel four times in a minute, so that the churn would turn at 48rpm, and it was estimated that 250lb of butter could be made in $1\frac{1}{2}$–2 hours in each of the two churns being driven. The horse had short chains from its collar to the ends of a substantial inverted-U-iron framework projecting below the wheel.

This idea was also used to drive a cider press, or else the simpler rotating arm of the sugar press was used, both with a horse moving a heavy stone roller around a circular trough at the centre.

A more portable arrangement had the rotating arm and axle on a base that could be transported and pegged, or otherwise fixed, to the ground. This turned a large bevel gear at the base, driving a smaller one on a take-off shaft, over which the horse had to step every time round (Fig 1B). For

Fig 1

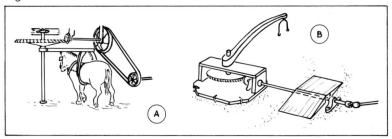

convenience the take-off shaft was in many pieces, joined with universal or flexible joints, and was taken over the ground to the job through many temporarily set up bearings.

The usual arrangement had one horse, but where more power was needed similar ideas were used for harnessing up to four horses. Such a portable assembly was used at the Great Exhibition of 1851, with four equally spaced horses going around, while a driver sat on a stool at the centre and went round with the roundabout. At the Exhibition this was used to drive Garratt's threshing machine. With the horses doing a steady 2mph an output of sixty bushels per hour was claimed.

As pieces of engineering some of these assemblies were comparatively inefficient, with much of the power produced by the horse lost on the way to the job, but the result was still very much better than could be achieved by men turning handles or otherwise labouring. Horse power used in this way overlapped well into the steam-engine era.

This set-up was used to work elevators and other static machinery in the field or yard, or the shaft might be taken into a barn to drive threshing machines, turnip cutters and other barn machinery. Many machines that became motorised had their birth in the days of horse power and needed little adaption.

Donkeys and mules seemed more amenable to this sort of work than horses. Another way of using them was to let them walk inside a vertical, drum-like wheel. This was certainly less likely to make them giddy. Power could be taken off in a similar way to the vertical shaft, but this arrangement was more suitable for a permanent assembly, where power was always wanted in one place, such as in a barn. Examples can still be found where a man or animal walked the wheel to draw water from a deep well.

Similar smaller wheels walked by dogs were used to operate a spit and keep the meat turning as it cooked in front of the farmhouse fire and traces of these may also still be seen.

Wheeled transport

Carts and waggons of the counties of England could more than fill a book themselves, but there are certain special features and broadly defined characteristics that help to identify types. Conditions on farms were often such that wheeled vehicles could not be used and anything to be transported had to be carried by man or animal. The use of any vehicle on uneven, soft, or rutted land depended mostly on the strength of the wheels. The British wheelwright became a very highly skilled craftsman, but it was not until early medieval times that satisfactory wheeled vehicles came into much use on farms. Early wheels had been solid, either a section of a log, or built-up from flat pieces joined across each other. A spoked wheel was much better, but required more skill to make. Solid wheels may still be found in primitive countries.

In a typical spoked wheel, as it developed and was used on farms up to the end of the horse era, there was a large wooden hub (nave, stock) with spokes tenoned into it, and a rim made up of felloes (pronounced 'fellies') in sections held by an iron rim, either a series of strakes or the more satisfactory continuous hoop. The important thing was the 'dished' shape of the wheel (Fig 2A). As a horse pulls a cart there is a swinging action from side to side, thrusting against each hub alternately. A wheel made without dish would soon have its centre knocked out. On the unmade surfaces of local roads and lanes the wheels made ruts in the ground, so it was important that all local vehicles had the same distance between wheels if they were not to have difficulties with the ruts. There is a theory that when steam railways were being made, their early builders measured the distances between ruts in their own local lanes and settled on that as the width between the wheels of the railway lines. This was 4ft 8½in and it has remained the gauge used by most railways of the world. Discoverers of Neolithic lake villages, near Meare in Somerset, found guide rails for vehicles preserved in the peat, and these were also 4ft 8½in.

A further advantage of the dished wheel was to make the width between the wheels greater at the top, as the axle dipped down to make the lower part of the wheel upright. This allowed the body of the vehicle to have flared sides and a greater capacity.

A cart has two wheels and a waggon four. Early carts were simple platforms supported over an axle, with a pole for pulling, probably at first by a pair of oxen. Later carts had shafts for a horse. The name 'tumbril' was used as an alternative name for a cart and in some places was only applied to a tipping cart. A trap was the light two-wheeled vehicle that was the rural predecessor of the family car.

Farm carts were mostly unsprung. The axle (exbed) was often wood, although smith-made iron axles were used. The basic structure (which might now be called a chassis) consisted of substantial lengthwise sides

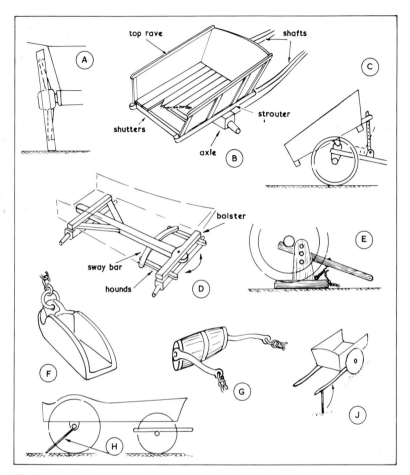

Fig 2

(soles), with crosswise shutters. The shafts continued the lines of the sides. The sides and front were built up from broad boards and sloped outwards. To facilitate emptying, the back of the cart was wider than the front and the boards forming the bottom were put in lengthwise, so that joints would not catch in shovels or other tools (Fig 2B).

In a tipping cart, the body, on a stout central crosswise member,

Plate 2   Gloucestershire waggon at Oxford City & County Museum, Woodstock

pivotted on supports over the axle line. The shafts were attached by framing to the axle, and some sort of device (tipstick) was arranged to lock the body down or secure it at various angles (Fig 2C).

Carts had to serve many purposes, so extending pieces for the sides, extra frames (ladders) to contain straw and similar parts might be provided. A cart was much cheaper than a waggon and had advantages in being more easily turned and usually needing only one horse. As roads improved, carts were more lightly constructed, particularly on slimmer wheels, and were used as transport by many others besides farmers, but the heavy carts continued on farmland.

The use of waggons as general farm transport began much later than that of carts. The Dutch, who came to help clear the Fens (in Tudor times), are supposed to have brought waggons with them.

Waggons varied in their design according to local ideas, and these traditional designs continued well into the periods of wider travel and the coming of mechanization. Nearly all farm waggons have the front wheels smaller than those at the back. These are attached to a forecarriage, which can turn for steering, so that the smaller wheels allow a greater

lock before they come against the waggon side (Plate 2). In most waggons the two axles are linked by a stout coupling pole, braced to the rear
axle, but on which the front axle pivots. This axle is braced by hounds
and a sway bar that rubs under the coupling pole (Fig 2D). Usually the
pull came directly on the undercarriage and the body of the waggon was
separate, acting as a container that was not stressed when a team of
horses pulled the waggon.

It is in the design of the body that waggons vary. In the days when all
wood was split and cut by hand, much thought went into selecting natural curves to suit parts of a waggon, so the grain followed the shape,
imparting strength, and reducing the amount of shaping needed. In
some local designs this is seen in the way a waggon body was waisted,
and the floor might curve upwards, to allow the front wheels to turn just
that little bit more without fouling. The alternative, in a few cases, was
for the front wheels to be small enough, and the body high enough, for
the front wheels to pass under the cart when turning, but very small
wheels meant a low axle which might foul the ground.

Later carts show adaption to the greater use of straight-cut planks
which came with power sawing. These were less beautiful and might be
slightly less effective for their purpose, but cost was considerably less. As
wood is not a durable material, it is mainly waggons showing this influence that have survived. Some of the waggons with straighter lines have
notches to allow wheels a fuller lock. Wheelwrights did quite a lot of
decorating in their work, but much of this had the additional purpose of
lightening the construction without weakening it. Bevelled edges on
any object at all are known to cabinetmakers as 'waggon bevelling'.

In most parts of the country the top of the body was higher than the top
of the rear wheels, but in the southern Midlands and the West Country,
the wheels were higher and the overhanging rave was curved over the
wheel (hoop raved), giving what must have been the most attractive form
of waggon. Of these the Woodstock waggon (Cotswolds, north of
Oxford) was claimed to be the most elegant.

Fortunately, the forests of Britain produced woods with sufficiently
varied characteristics to suit the parts of a waggon, with ash for shafts
and other parts requiring spring, oak and elm for hubs and many structural sections. Beech also found a place, particularly for more precise
parts. Softwoods, either home-grown or imported, had little place in
waggons and carts.

The wheel hubs were lined completely or partly with an iron sleeve or box, whether the axle was iron or wood. Such a bearing needed frequent greasing. This meant jacking up the axle to remove the wheel. Most cart jacks relied on a lever action (Fig 2E), although some later ones were more like the modern screw-type car jack.

Although coaches and some other road vehicles had brakes that worked by pressing against the wheels, the only braking system for a farm waggon was a shoe (skid pan, drug, bat) on a chain, that went under the wheel and skidded along the ground (Fig 2F). An alternative was to lock the wheel with a chain, but this obviously caused wear on the iron tyre on a hard surface. A different problem came when going uphill and there was a risk of the horse being unable to prevent the cart or waggon running back. To prevent this, a 'roller scotch' (Fig 2G) was trailed close behind one wheel. (This has given rise to the saying 'to scotch' something, meaning to stop it.) Another precaution was to trail a dogstick under an axletree. If the waggon tended to slip back, this dug into the ground and acted as a strut (Fig 2H). Shafts often carried a prop-stick (Fig 2J)—let down to take the weight off the horse when stationary.

Carts and waggons used on farms did not have a seat for the driver. He walked, rode one of the horses, sat on one shaft or found a place on the load.

With the coming of steam engines and internal-combustion-engined tractors, many carts and waggons were adapted to be towed by them, but they were not designed for speeds much higher than walking pace and they suffered from this treatment. Until the coming of inflatable rubber tyres, waggons were in a rather undecided state, but with the change to rubber-tyred wheels, trailers more in keeping with tractors came into use.

Steam power

Early steam engines merely produced an up and down motion, which could be used to work a pump. In 1781, James Watt, who had experimented with steam engines as early as 1754, made an engine that converted this movement to a rotating action. Richard Trevithick, a Cornishman, made an engine into a locomotive and ran it in South Wales in 1803, thus marking the start of something that could have application in agriculture.

Plate 3   Threshing at the rick, with a threshing box driven by an 8hp Ransome's portable
steam engine, about 1870

In a steam engine a fire is used to heat water which turns into steam. This steam is fed under pressure via a valve to a cylinder where it pushes a piston along the cylinder and is then released via another valve. In most steam engines the steam is fed alternately to opposite sides of the piston so that it moves both ways. A connecting rod from the piston goes to a crank on a shaft carrying a flywheel. In this way the reciprocating motion is changed to rotation and the heavy flywheel smooths out some of the jerkiness of the reciprocating piston. There may be a second cylinder. The valves to admit and release steam from the cylinders operated by linkage from the crankshaft.

Steam engines, as they developed for use on the land, worked in this way. They were external-combustion engines, as distinct from the later internal-combustion engines, where firing occurred in the cylinders. There were many improvements to the boilers, with water and fire tubes.

25

Although a steam engine can carry a certain amount of water and coal, it has to be fed frequently with them.

The mole plough had been thought of before the steam power for hauling it became practicable. Mole ploughs were winched across the field by hand. A patent for using a steam engine for doing this was granted early in the nineteenth century, and set the pattern for cultivating ploughs, hauled across a field by a stationary steam engine, but it was not until the middle of the century that steam ploughing seems to have become feasible.

Steam engines, first brought onto the land for ploughing and other purposes, were portable in the sense that they could be hauled by horses, but they were not self-propelling (Plate 3). A team of horses was needed to move one, so for steam ploughing there had to be horses and men available to move the engines after ploughing each set of furrows. Such engines could also provide power for elevators, and they made successful large threshing machines possible. Despite its cumbersome nature, the slow-turning high-torque output of a steam engine could be used to operate equipment and implements outside the scope of any team of horses, and work at a much higher rate than any other methods available, in situations where the need was large enough to justify positioning the engine and its gear.

It was not long before steam engines became self-propelling, with the engine driving the rear wheels and the front ones steered by a worm and pinion arrangement. Earlier steering was by the horse-drawn forward carriage.

Later engines became lighter, but the functioning of a steam engine necessitates a fairly heavy construction, so there was a limit. Consequently steam engines, for travelling over the land towing implements or having them built-in, did not have universal application. The wheels had broad surfaces to minimize sinking, but the passage of a steam engine had a compressing effect on the soil which was often undesirable.

Early steam engines had a variety of layouts, but portable engines for farm use soon took the familiar form of one or two cylinders mounted above a horizontal boiler, with the chimney at the forward end and the crankshaft with a large flywheel at the rear end (Fig 3A). If the engine was self-propelling, drive was to the large rear wheels. The driver rode on a platform behind the boiler and firebox, with the coal and water tanks behind him. He steered the front wheels with worm gear and cables. The

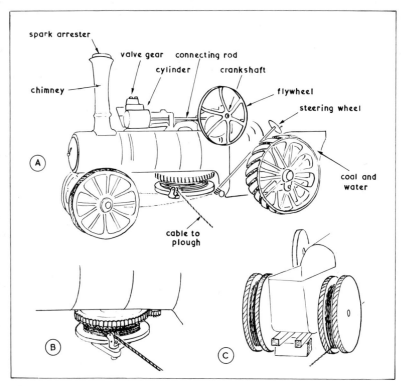

Fig 3

very high ratio of the steering wheel only permitted slow turning, but at a speed less than walking pace it was adequate.

At first steam engines did their work through what the modern tractor driver refers to as 'PTO' (power take-off). A belt from the flywheel, or a pulley beside it, drove threshing and other machines. For ploughing, which was one of the early successes of steam (see Chapter 4), the cable was usually wound on a horizontal drum beneath the engine, driven by bevel gearing from the engine shaft (Fig 3B). For some systems there were two drums. The most popular system of ploughing used two engines at opposite sides of the field. This was obviously more expensive in first and running costs, but gained in speed and convenience, so was practicable for contractors.

27

Vertical winding drums were used by some manufacturers. Another way of winding the plough cable was to use deep, broad grooves in the main, road wheels. In use the engine had to be chocked up so that these wheels were clear of the ground (Fig 3C).

Steam tractors came into use for towing loads and either pulling ploughs or other implements, or actually having the implements bolted on, in the same way as is done today with internal-combustion tractors. Although progress had been made in building these steam tractors lighter than the ploughing engines, they were still very heavy and this limited their use to land where there was no risk of sinking in. The introduction of high-pressure boilers in the 1850s did much to lighten engines. Towards the end of the steam-farming era, steam tractors bore family likenesses to the operative ends of the steam lorries still in use on British roads in the 1920s.

A well-known British make of steam tractor was Mann's 'Agric', used mostly between 1915 and 1925. What was believed to be the last steam tractor built was the Garrett 'Suffolk Punch', about 1925, when four were completed. Horses and steam engines existed together, but both found their work increasingly taken over by the internal-combustion-engine tractors after World War I.

Smaller, stationary steam engines were used successfully to drive yard and barn machinery. The larger, portable engines were also brought in for this work when not required in the field. However, stationary oil engines became available for farm use before the arrival of tractors, so these much less troublesome engines soon took over. With any steam engine there has to be a period of warming and raising a head of steam before any work can be done. This period depended on the type of boiler and the skill of the attendant, but in general anything up to an hour had to be allowed for preparation.

The development of the steam engine coincided with a great enthusiasm for engineering generally. Maybe the power provided by the steam engine, greater than anything known before, gave the impetus needed to tackle constructional and mechanical work in a more scientific way. In agriculture, this meant implements were improved, being stronger and able to work more efficiently. All the incidentals like shafts, pulleys, belts and other means of transferring power became better engineered.

These improvements were additional and complementary to the steam engine, but they all added up to a situation where agriculture and industry were ready for the internal-combustion engine. If this had come without the steam engine there might not have been the knowledge and equipment able to benefit immediately from it.

The output of steam and internal-combustion engines is quoted as horsepower. Unfortunately, there have been many ways of arriving at a horsepower rating and in these days when the high figures quoted for some motors might be interpreted as wishful thinking, the more matter-of-fact horsepowers of early steam engines seem very low—5 or 10hp was usual. It is said that the work output of an average horse was measured, then doubled, so that engine designers could not be accused of making extravagant claims when they quoted horsepower. The figure arrived at was 33,000 foot-pounds per minute—the ability to lift 33,000lb through 1ft in 1 minute, or a proportionately smaller amount higher or quicker. Brake horsepower is usually the preferred figure, being measured by an instrument at the shaft where the power is taken from the engine.

Relating the work of a machine to that of a horse cannot be very precise. A horse gets tired, so its work does not remain consistent. A two-engine ploughing rig could work a six-furrow balance plough and cover perhaps fourteen acres in a day. A single-furrow horse plough might not manage much over an acre in the same time. The man, as well as the horse, walked a long way. Although steam ploughing meant a considerable investment in equipment and at least three men were directly involved, with many others supplementing them, on balance, steam was showing the way and leading to the mechanized farming that has been carried to its present stage by the internal-combustion engine.

CHAPTER 3

The Coming of Tractors

Internal-combustion engines are broadly divided into those that have spark ignition and those that use compression ignition. Engines using gas, petrol (gasolene), paraffin (kerosene) and some other vapourising oils, use spark ignition. Compression-ignition engines burn heavy oil. Nearly all petrol engines used in agriculture are four-stroke, running on what is called the 'Otto cycle'. The alternative two-stroke principle is rarely found, except in light market-garden equipment and lawn mowers.

Spark ignition is provided at a spark plug, which has changed little since the early days. In each cylinder, which contains a piston connected to a crankshaft in the same way as in a steam engine, there are inlet and exhaust valves. On the induction stroke (Fig 4A) a mixture of air and gas is sucked in through the open inlet valve. On the compression stroke (Fig 4B) the fuel mixture is compressed with both valves closed. On the expansion stroke a spark occurs igniting the compressed mixture, which expands and forces the piston down (Fig 4C). On the exhaust stroke (Fig 4D) the piston returns and the burned mixture is forced out through the open exhaust valve. The piston is then in position to start the cycle again.

At one time electricity for the spark was provided by a magneto, but now it is more likely to be provided by a high-tension coil. The correct fuel/air mixture is provided through the carburettor. Many engines have four cylinders, but there can be any number. By spacing the sequence of power strokes evenly, smoother running is achieved. A flywheel further controls this. Today electric starting is general, but most early engines relied on hand cranking.

Experiments with volatile fuels had been made by several inventors around the turn of the nineteenth century, but the internal-combustion

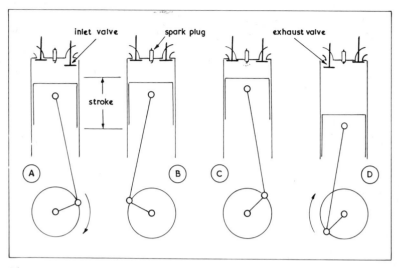

Fig 4

engine, which developed into the motor we know, started when a successful gas engine was offered for sale by Lenior in 1860. Soon after that came an engine by Otto, after whom the four-stroke cycle is named. Oil engines followed from several inventors. Daimler patented a high-speed petrol engine in 1883, but the credit for the first reliable oil engine is given to Priestman in 1888. There is little difference between an engine designed to run on town gas and one intended for petrol or vapourising oil. In many cases one can be adapted to run on the other fuel as the principle is the same.

Compression-ignition engines came later. The first patents were granted to Ackroyd-Stuart in 1890. Diesel proposed a compression-ignition engine in 1893. The first successful diesel engine ran in 1897. Diesel's name is now the usual one for any compression-ignition engine.

A compression-ignition engine needs to be stronger than a petrol one, so diesel engines tend to be more massive, although in recent years they have become lighter. A very high compression generates heat. The comparatively unvolatile fuel is sprayed in just before the air in the cylinder has reached maximum compression, then auto-ignition takes place. Precise metering of the fuel from the pump is important. For a long

31

time diesel engines were only used for large equipment and heavy vehicles because of the size and weight of the engines, but the newer and lighter versions are of a suitable size for tractors and diesel tractors are becoming more common than those running on petrol.

At one time other fuels were used instead of petrol, for economy. Vapourising oil (paraffin, kerosene) was used after a spark-ignition engine had been warmed up on petrol. This gave slightly lower power, but the fuel was cheaper. With the coming of diesel-powered tractors, vapourising oil as an agricultural fuel is being used much less.

Early agricultural engines

The first internal-combustion engines used on farms were stationary oil engines used to drive barn machinery. Other versions could be towed by horses in the same way as early steam engines. Successful oil engines came on the market in 1890. These were produced by Herbert Stuart and Charles Binney and taken up by Richard Hornsby and Sons, a firm still well known.

These engines ran on lamp oil, which had to be heated with a blowlamp for starting, but once running they worked indefinitely with the lamp removed. Hornsby sold these engines for some time, and it was estimated that there were 13,000 stationary oil engines in use on farms by 1908, but he failed to develop them into tractors.

In 1902 a young experimenter named Dan Albone produced the Ivel tractor. This had three wheels, with a single small one at the front and two large slatted ones at the back. Steering was by a wheel, with cables to a pulley on a vertical shaft over the front wheel. Most of the weight was concentrated over the rear wheels. The engine ran on paraffin and had two opposed cylinders 4 7/16in bore and 5in stroke. At the carburettor, engine suction drew air through a reservoir of fuel. Inlet and outlet valves were automatic and spring-loaded. Coil ignition was used. There was no radiator, but water circulated through a thirty-gallon tank. Drive was through multiple chains and sprockets and the rear axle had a differential. The simple and massive clutch gave one forward and one reverse speed. Albone died in 1906, but the company continued until 1921 and produced nearly a thousand tractors.

Ransomes, Sims & Jefferies, already famous for their agricultural implements, produced a four-cylinder 20hp tractor, which was very like the cars then being produced. Although it did all that was claimed for it, the firm soon withdrew the tractor and Ransomes did not come back to the tractor market for ten years, when they offered a single-cylinder 35hp ten-ton self-propelling oil engine, that looked remarkably like a steam traction engine, with the exhaust going up a chimney. This was more of a road vehicle as it was too heavy to travel over most cultivated land.

Tractors of that time with oil or petrol engines tended to be built as cars for use on the land, or were laid out in a similar way to steam engines. The tractor form general today was slow to evolve. A lighter and therefore more suitable tractor for drawing implements was produced by a firm named Sharp, of York. This weighed 15cwt and was claimed to haul two tons across a field, and to move a three-furrow plough at 5mph. This was obviously what was wanted and pointed the way development had to go.

In 1910 the first officially-sponsored tractor trials were held in England by the Royal Agricultural Society. This was open to steam as well as other tractors, and three steam tractors competed against two Ivel tractors and two Universals by Saunderson & Mills. The judges were inconclusive, but tended to favour steam, possibly reflecting the conservatism of farmers generally.

American developments

Parallel developments were taking place in America. There was obviously some exchange of ideas, but generally progress on one side of the Atlantic went on without much reference to the other. In America, where the size of the country meant that much farming was over vast areas, there was more incentive to develop efficient mechanical power to replace the horse and steam. An early American portable steam engine had been the first by the J. I. Case Co in 1869. This was similar to British steam engines of the period. Case went on to become famous for tractors and implements. Their first tractor had an opposed pair of cylinders, but suffered from unsatisfactory ignition.

In 1902 C. W. Hart and C. H. Parr, of Charles City, Iowa, built a six-ton tractor powered by a two-cylinder four-stroke kerosene engine. This, and others that followed from these makers, looked very like a steam engine. The most prominent feature was a large stack at the front. It had a rugged transmission which was claimed to be able to cope with the heavy strain imposed by ploughing. Iowa is a state where there would then have been the need for cultivating large tracts of hard land, which could take the heavy weight of this tractor. The Hart-Parr Co is supposed to be the first to use the word 'tractor'.

In 1899 S. S. Morton built a successful tractor and took out patents, in 1902 and 1903, which were bought by the Ohio Manufacturing Co. International Harvester marketed them in 1905, but in 1906 they started building their own 10hp tractor with reverse gear and a friction drive. This was the start of the firm which has now spread to many parts of the world. In 1910 they built the first of their series of Mogul tractors. This had a two-cylinder 45hp engine, low tension make-and-break ignition and spur gear transmission.

Probably the most prophetic design was a tractor designed by J. S. Holmes in Beloit, Wisconsin, for use on his own farm in 1906/7. Photographs in the Smithsonian Institute, Washington, show the tractor with iron wheels, the same size back and front, the driver on a small, shaped seat and the engine, fuel tank and, apparently, a water tank that served the purpose later taken by a radiator, all in an assembly that, at a distance, had a profile very similar to a modern tractor. Photographs show the tractor hauling a heavy trailer, and ploughing. A second tractor built by Holmes could take the front end of a waggon at the back, forming what must have been one of the first articulated vehicles or semi-trailers. Holmes did not go into commercial production of these tractors.

One of the successful early American tractors was the Waterloo Boy. This started with experiments by John Froelich in 1902 in Iowa. He built a gasolene (petrol) tractor and, further experiments and many models later, came up to the Waterloo Boy Model R in 1914, as a production tractor. By 1918 the Model N had a two-cylinder kerosene engine of 25hp with two forward gears and a reverse. The Waterloo company was taken over by the John Deere Co and continued to produce Waterloo Boy tractors until 1923.

After World War I

It was World War I that gave a great impetus to the development of tractors, and farmers who had resisted were compelled to accept them if they were to keep pace with demands for food production. With farm workers being called up and horses commandeered for war work, mechanical power on the land was becoming increasingly important. In the years leading up to the war there had been much progress, and tractors were becoming more reliable and easier to manage with all the inventions, improvements and modifications that the many manufacturers were incorporating.

Henry Ford was busy in America producing his Model T and other cars, but he had also turned his attention to tractors. Conversion kits became available to change a Ford car into a tractor and Henry Ford made his first tractor in 1907. Constructionally it was very much a mixture of parts taken from other machines, but it had a 24hp four-cylinder engine. In 1911 a British branch started assembling Model T Ford cars in Britain, under Percival L. Perry (later Lord Perry), and this British connection was the vital link that brought Fordson tractors, amongst other products, to the aid of Britain during World War I.

By 1917 the German submarine blockade was making itself felt and Britain's food reserves were very low. At that time Henry Ford had built a large factory at Baton Rouge, USA, for the production of Fordson tractors. With his production-line assembly he was building enormous numbers—sales passed 100,000 in 1920. Prime Minister Lloyd George arranged for 5,000 of the tractors produced there to be delivered within three months. Besides helping to solve the British farming problems of the time, the impact of the introduction of this large number of tractors on agriculture did much to influence the attitude of the workers concerned, so they were more receptive to their use and further mechanization in the postwar years.

A British tractor that seemed ahead of its time, but which never achieved large production, was the Ideal, produced from 1912 to 1918. This had a four-cylinder 35hp Dorman petrol engine and weighed $4\frac{1}{2}$ tons. It could plough four furrows and had a power lift for the plough—probably the first British tractor to have this. There was a belt pulley and a drive to a grass cutter. It had retracting spuds (slats) across the rear wheels. These could have varying projection to suit land conditions and could be fully withdrawn for travel on the road. Unfortunately stones and mud jamming the grooves made this idea unsatisfactory.

In all early tractors the wheels were iron and it soon became general for the rear wheels to be larger than those in front. To carry the weight without sinking in, the wheels were always broad. Front wheels might have ridges or grooves around their circumference to aid steering, but the rear wheels were given crosswise slats (spuds), similar to those used on steam engines, to provide a grip. Some wheels had extra spike rings that could be attached to provide a grip.

The problem of getting a grip on a very bad surface had been met quite early in tractor production by making an endless belt with slats around front and rear wheels. These are rather cumbersomely described officially as 'track-laying vehicles steered by their own tracks'. The more usual name today for this sort of tractor is a 'crawler'. The word 'Caterpiller' applied to this kind of tractor is actually the trade name for the products of a particular company. Crawlers probably made their name by their adaption to fighting tanks during World War I.

There were and still are regulations that make taking iron-wheeled or crawler tractors on the public road difficult. This reduced the usefulness of these tractors, and the problem continued until the coming of satisfactory pneumatic tyres.

It was the Firestone Tyre & Rubber Co in the USA which made the first, successful, tractor pneumatic tyres. These were adopted by Allis-Chalmers for their tractors in 1932 and they soon became usual on all tractors. Solid rubber tyres had been used earlier for tractors off the land, but they were unsuitable for agriculture. At one stage there were comparatively narrow pneumatic tyres suitable for use on the road, which could have iron rims added for use on the land. This is still done for heavy loads. Narrow tyres have almost completely given way to the very large tyres now commonly used, with large rubber slats to provide a grip, that are equally suitable for road or land.

Some quite early tractors had provision for a power take-off in the form of a pulley for a belt, so that the stationary tractor could be used as the power source for a machine such as a thresher, but it was the International Harvester Co who perfected a power take-off mechanism on their tractors in 1918, so that equipment mounted or being towed could be driven when the tractor was moving. This meant that implements that had taken their power from the ground wheels, which were subject to slipping, could be operated by the tractor engine, giving more positive and regular speeds, with control of speed possible, independent of how fast the road wheels were turning. This facility soon became usual on most tractors.

Towards lighter tractors

The name 'tractor' had become firmly established by the end of World War I. With the influence of Ford and others, tractors had become much lighter. People other than farmers or engineers with steam experience had become involved, and tractors were produced without the great weight which had been unavoidable with steam. Early Ford tractors weighed about 1cwt for each 1hp. Many tractors by other makers were three or four times as heavy.

Although a case could be made for the larger and heavier tractors in use on the vast farm areas of America, the need in Britain was for lighter models. Two heavy tractors that found their way into Britain, mainly to aid in agriculture during World War I, were the International Harvester Mogul and Titan. Their construction was so substantial that some of these machines are still in existence and in working order. The first Mogul was produced in 1910. Other models followed, but it was the

Plate 4   A 1916 International Harvester 'Titan' tractor, one of the large number of these American tractors that came to Britain during World War I

8–16, which had a single-cylinder 16hp engine, that came to Britain. This weighed about 2¼ tons, which was even then considerably less than earlier models. By the time production finished in 1919 over 20,000 Moguls had been produced.

However, it was the International Harvester Titan that came to Britain in greater numbers and is more likely still to be found (Plate 4). There were several models, but the 1914 version with its twin-cylinder paraffin engine produced about 20hp and weighed 2½ tons. A 1915 model had a four-cylinder double-twin engine and weighed nearly 4 tons. Many Titans had a large cooling water-cylinder in the front and a fly-wheel, so they looked like steam engines without chimneys. By the time Titan production ceased in 1924, over 60,000 had been produced.

In 1919 an attempt was made to assemble as many tractors as possible for trials at South Carlton, Lincoln, so that farmers could see all that were available after the war. A great many firms participated, but many appear to have had only short lives. By then Austin, the car firm, had produced an economically priced 25hp 28cwt tractor. Ruston and Hornsby produced their British Wallis with a 23hp four-cylinder paraffin engine. The Royal Agricultural Society followed with trials in 1920 at

Plate 5   American 'Farmall' three-wheel tractor, popular between the wars, with the solid-rubber-tyred front wheel, controlled by a worm drive over the engine. The narrow rear wheels have prominent spuds to provide a grip

Plate 6   British Anzani 'Iron Horse' cultivator, a walking tractor, which could take a
plough and other attachments

Aisthorpe, near Lincoln, a noteworthy feature of these trials being the
fact that cable ploughing was still considered a reasonable alternative to
towing the plough with a tractor. Fowlers, famous for their steam
ploughing-engines, showed their motor-cable plough engines. They
were petrol engines, 46hp, four-cylinder and had electric starting. Appearance was very much the same as a steam engine, and the method of
working followed the earlier Fowler practice with steam engines.

  J. & H. McLaren, of Leeds, produced their Motor Windlass. This had
a diesel 60hp motor, with an 8hp auxiliary petrol motor for starting. The
cable-ploughing winding drum was vertical. Several winding speeds
were possible. Like the Fowler engine, the machine was quite heavy and
self-propelling. The McLaren machine proved more popular and many
of these continued in use in Britain and abroad for most of the be-
tween-war years.

  Tractors generally are regarded as an alternative to the horse and for a

39

long time were treated as such, with implements towed behind and often tended by a second man. The logical development was to build in the implement by designing it to attach direct to the tractor and this increasingly became the way the use of tractors developed.

One of the earliest tractors to have such a power take-off for working attached implements was the McCormick-Deering 10–20 by International Harvester. This was one of the 'Farmall' series of tractors (Plate 5). An earlier version was the 15–30 started in 1921. The figures indicate the horsepower at the drawbar (15) and via the belt (30). The 10–20 was lightest at $1\frac{1}{2}$ tons. The other was $2\frac{1}{2}$ tons. Both had four-cylinder vertical engines, with cylinders that could be replaced, and ran on paraffin after starting on petrol. The general appearance of these two tractors was very similar to modern machines. Production continued through a peak in 1929 and did not cease until 1940 when sales of 10–20 had exceeded 215,000. Tractors that had been discarded during the 1930s came back to do a good job during World War II.

Another development alongside that of tractors was the mechanization of the implement itself. This was described as a 'walking tractor'. For instance, ploughs were designed with two large tractor-type wheels at the front, driven by an engine. The driver walked behind and controlled the plough in much the same way he did when horse-ploughing (Plate 6). The use of an engine allowed multi-furrow ploughs to be employed. The same idea was used for other tilling implements, but experience has shown that for farm work it is better to have implements with a full-size tractor.

It was in 1920 that Harry Ferguson designed his first three-point linkage for attaching implements to a tractor. Implements drawn by a horse had the pull applied fairly high from the animal's neck and shoulders. With the coming of tractors they merely replaced the horse and the trailed implement was still pulled from quite high up. With the three-point linkage the implement became integrated with the tractor. The bottom two links drew the implement forward, the upper link applied force from the action of the implement to the tractor; the effect being to add weight, and therefore better traction, to the tractor's rear wheels.

Tractors were made to Ferguson's designs in Huddersfield and, by 1935, he had perfected the three-point linkage system with a complementary hydraulic system suitable for a wide variety of farm implements. In 1938 Harry Ferguson and Henry Ford made a verbal agreement by which Ferguson was to stop production of tractors. Instead, they were to be made by Henry Ford in America. Relations between the two companies deteriorated. In 1945 the building of Ferguson tractors started in Coventry. In the famous lawsuit that followed Ferguson was awarded over £3 million from Ford for royalties due on his patents.

In 1953 Ferguson amalgamated with Massey-Harris of Toronto, Canada. Massey had started making agricultural machinery in 1864. Harris had been doing similar work and they came together towards the end of the nineteenth century. After 1953, tractors and other equipment were sold, as they still are, under the Massey-Ferguson name.

Tractor progress since the 1920s has paralleled that of motor cars. Engines have been the same in many cases and transmission has become similar to that of cars. Gear boxes are also very similar, and electric starting has become normal. Early tractors had few, if any, instruments while modern tractors have dashboards equipped at least as well as the family car. Because most tractors are taken on the road they have signals and lights, and the steering wheel is also car-type.

Many modern tractors are large and complex machines in themselves

Plate 7 Modern tractors may have a variety of attachments. This Leyland 270 carried an LK liquid nitrogen fertilizer injection system. Note the extensions to the rubber-tyred wheels

with provision for further complication by the addition of many accessories (Plate 7). Four-wheel drive is available. This, with the added grip provided by modern tyres, makes possible much of the work that would otherwise have been done by crawlers, without the restrictions that are imposed on a crawler when taken on the road. Hydraulics play a larger part in the mechanisms of a tractor and they can be used to operate implements attached, towed or fitted underneath. Bucket and fork lifts, front and back, are commonly used.

Driver comfort and safety have received much attention. Supporting arches to protect the driver if the machine turns over are provided. Controls are better arranged and often inside soundproof cabs with comfortable seating. Apart from safety these facilities enable a worker to give his maximum output.

CHAPTER 4

Ploughing

From the time of the digging stick, whether pointed or spadelike, primitive man tried to speed and ease the job of tilling the soil by the application of levers and simple mechanization. A variation on the spade or stick that just about doubled the speed of digging was the 'cas chrom', or foot plough, in which a lever was applied to the stick and usually a step or leg for foot pressure added. This may have been a suitably shaped branch or a built-up tool (Fig 5A), and the end might have been strengthened with horn. Steel-edged modern versions have been in use in

Fig 5

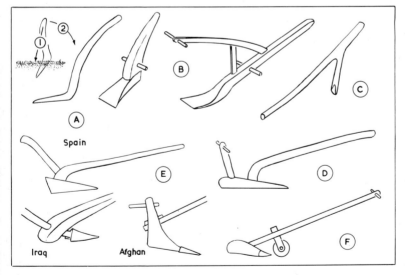

Spain

Iraq   Afghan

the Scottish Western Highlands and Islands until recent times. Another version (Fig 5B), credited to China and elsewhere, has a forward handle so a helper can assist with the rocking action.

In use the cas chrom is tilted forward and the edge driven into the soil with a foot on the peg. The handle is levered backwards to make the point lift up through the surface and the worker moves back to repeat the action. Of course, this merely breaks up the soil without any ordered furrows, although a skilled worker would turn most of the soil one way each time. On new ground the surface turf and weeds would first be scraped off with a breast plough (Fig 17P, p 89).

Another variation on the digging stick was an adaption to pull it, making something with an action more like a modern plough. This might have been a natural branch (Fig 5C) or a built-up tool. A further step was a handle (Fig 5D), so that one worker followed and guided the tool while others, or animals, pulled it. Such ploughs, in which the idea has not been taken much further, are still to be found in parts of Spain as well as more primitive countries (Fig 5E). Hand ploughs (Fig 5F) have also been used in more recent times in orchards and other restricted places.

All early ploughs were merely scratching tools, lifting and separating the soil to either side. This left unploughed strips between furrows, and one way of improving on this was to plough again across the first ploughing. There are Egyptian drawings showing many ploughs using the same principle, several having double handles, and most pulled by two oxen. The Greeks followed on with similar tools, all of which would have been of wood except for an iron share.

Roman ploughs were similar, but there are records of wheel ploughs and the use of something very like a modern coulter in front of the share during the days of the Roman Empire. The Romans brought their ploughs to Britain, but the British farmers, both before and after the Roman occupation, appear to have favoured large and heavy ploughs, pulled by as many as eight oxen.

Plough parts (Fig 6)

In order that differences between the enormous numbers of plough types can be appreciated, it may be helpful to know the names and purpose of the usual parts of a horse-drawn plough of more recent times.

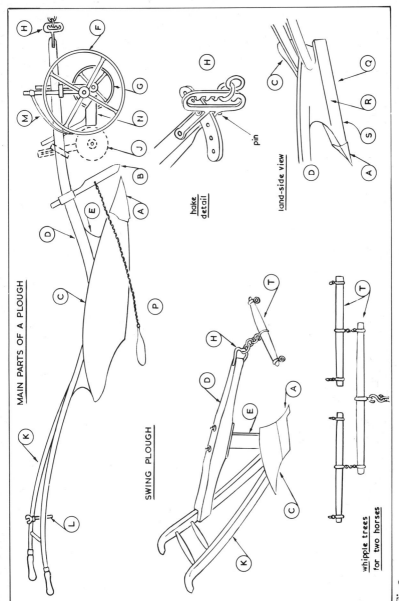

MAIN PARTS OF A PLOUGH

hake detail

pin

land-side view

SWING PLOUGH

whipple trees for two horses

Fig 6

Earlier ploughs may have lacked some of these parts and modern tractor-powered ploughs do not need all of them. Names varied between localities, but what seem to be the usual ones are given first.

A plough (plow—USA) without a pair of wheels is called a *swing plough*. Some swing ploughs have a small wheel, but this is to support the beam if it tilts down, or to run on the unploughed landside only. If there is a need to distinguish it, the more usual British model is called a *wheeled plough* (furrow plough, ridging plough). Nearly all ploughs are made of iron and steel, but there are still ploughs with wooden parts. So that a great many variations in assembly can be made, many parts are held together with metal straps and nuts and bolts. The expert ploughman makes many ad-

Plate 8   The action of ploughing—the horses provide the power through the whippletrees and hake, while the driver adjusts parts to suit the ground, then controls by bearing on the handles

46

justments before he is satisfied that he has his tool correctly set to suit the particular ground being worked. (Plate 8).

The share (sock) (Fig 6A) is the cutting part of the plough, which breaks into the soil. Early, crude wooden shares were called *windingboards*. Some later shares were wood with iron facing, but were more likely to be iron or steel. Cast-iron shares have been given self-sharpening properties by having the bottom harder than the top, so that the wearing away of the top leaves a sharp edge. Shares may be interchangeable so that one appropriate to the soil may be fitted. The angle of the share might also be adjustable.

The coulter (knife could, colter—USA) (Fig 6B) makes a vertical cut in

the soil immediately ahead of the share, although it can be positioned as preferred.

The mouldboard (breast, wing, turn-furrow) (Fig 6C) continues behind the share and is the most prominent part of the plough. Its purpose is to twist the furrow slice and turn it over. Its built-in twist is obviously important. This would have been arranged by eye and experience, but with increases in engineering knowledge, particularly in the nineteenth century, much thought was given to scientifically shaping the mouldboard. In nearly all ploughs the mouldboard is on the right of the plough (Plate 9). Mouldboards were first made of wood, they were then covered with iron, while later ones were made of steel. A type of three-ply steel has been used, with a tough soft-steel centre, providing strength for the harder wearing, but more brittle, hardened steel on the outside.

The main structural part of a plough is the *beam* (Fig 6D). All other parts are attached to it and it needs considerable strength to withstand the varying loads put on it. Consequently, in early wooden ploughs the beam was quite massive. Thrust from the share and mouldboard had to be transferred to the beam through *the body* (frog or standard) (Fig 6E).

In a wheeled plough the two wheels are on stub axles attached to vertical square rods which can be adjusted in height or width, or arranged,

Plate 9   The action of a plough—the furrow wheel follows the last furrow, the coulter cuts vertically, then the share cuts the sod which is turned by the mouldboard

one slightly ahead of the other, on a crossbar which fixes to the beam and can be located in several positions on it. The larger *furrow wheel* (Fig 6F) travels in the bottom of the furrow. The smaller wheel travels on top of the unploughed side. This is called the *land wheel* (Fig 6G). (A plough-man refers to the unploughed part as landside.) Differences in the height of the two wheels is the main control of depth of cut.

The pull from the animals is taken via the *hake* (muzzle, bridle, clevis—USA) (Fig 6H). In its simplest form this is merely a fixed attachment, but vertical, horizontal, or both, adjustments are usual. Horizontal adjustment, usually with a pin in a quadrant, could be used to correct any tendency of the plough to pull to one side. Similarly, altering the chain height could affect the working depth of the plough in relation to the pull.

A *disc coulter* (Fig 6J) is an alternative or addition to the *knife coulter* and can be adjusted and positioned in the same way. The disc has a stout centre, tapering to a cutting edge at the circumference. It is of Dutch origin and particularly good at cutting through turf. *Wheel coulters* have continued on tractor ploughs.

The ploughman followed the plough while holding handles at the ends of two *stilts* (USA—plow neck or plow tail) (Fig 6K). These are usually bolted to the beam and are braced together, as well as sometimes to a point further forward on the beam. The ploughman may put considerable loads on the stilts to control the plough, particularly when he tilts it onto one wheel at the end of a bout of work to release the share from the soil in readiness for turning round. There is even more load on the handles of a swing plough, where there is no benefit from the control of depth given by the wheels. This means fairly long stilts to provide leverage for the man, with a fairly short projection of the beam forward, to minimise the leverage due to the horse. The swing plough shown on p 45 was sketched from a pioneer plough in Utah, USA.

At least one *spanner* (Fig 6L) is carried in a socket, so that it is to hand for making adjustments. Some are combination tools, with a hammer head at one end. A tool like a short-handled hoe may be carried for removing mud from the mouldboard and other parts of the plough.

A wheel clogged with mud might be difficult to pull, and a build-up of mud could affect the depth of cut, so some ploughs have a *wheel scraper* (Fig 6M) at each side, mounted so as to keep the rims clear.

A *skim coulter* (Fig 6N) is a refinement not always provided or used. It acts like a small plough and is set ahead of the knife or wheel coulter at a suitable height, to cut and turn a small slice near the surface at the land-side edge of what will be the main slice. Its purpose is to turn over surface growth, weeds and other rubbish, then, as the main furrow slice is made, this is buried, leaving a clean appearance to the *comb* (apex, arris) of the furrow.

Sometimes there is a drag chain towing a heavy weight, called a *boat* or *seamer* (Fig 6P). This pulls along the edge of the furrow slice, breaking off growth so that it will be buried, and smoothing the edge of the furrow.

The action of ploughing puts a considerable thrust against the side of the plough opposite to the mouldboard, to be transferred to the landside of the furrow. This is resisted by a substantial *steel slade* (ground wrest) (Fig 6Q), usually of L-section, with its vertical part being the landside (Fig 6R), and the bottom part called the *sole* (Fig 6S), which slides along the bottom of the furrow and takes some of the weight off the plough.

The *chain traces* (draught chains, draw chains), on which the horse pulls, are kept clear of the horse's body by being attached to *whippletrees* (whiffletrees, swing bars, singletrees, swingletrees, swingle bars, strad-sticks, badikins) (Fig 6T). If there are two horses, the centre of each whippletree is linked to the ends of another (*doubletree*) which has its centre linked to the *plough hake*. If three horses are used, particularly with a double-furrow plough, there has to be a very wide and strong whip-pletree next to the plough. Simple whippletrees are ash wood with iron hands, but more heavily-loaded ones are iron or steel.

Western ploughs

Early ploughs, which developed from the primitive scratching tools, were built mainly of wood, with iron used at the points of greatest wear, if available. As wood is not sufficiently durable, specimens of early ploughs have not survived, but there are drawings from the early days, on tapestries and elsewhere, which help give a picture of very slow development.

Cumbersome and crude ploughs, but showing some understanding of the principles involved, were in use before the Norman Conquest in 1066, and continued with only slight improvements up to the eighteenth century. At this time there were improvements in design, mostly covered by patents, and iron began to take over from wood for the main structural parts as well as the points of wear. In the nineteenth century, plough manufacture became industrialised and the smaller number of manufac-

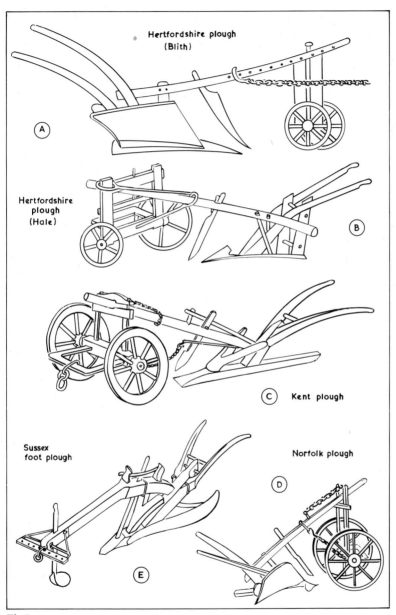

Fig 7

turers produced more effective and scientific ploughs, which became available to a larger number of farmers. This continued into the twentieth century, with an acceptance of mechanization and alterations to suit power farming.

More positive records are in books, which also show the thought that was, at last, being given to better plough design. In *The English Improver Improved* of 1653 Walter Blith shows the Hertfordshire plough. This was the general form of plough used throughout the Midlands of England (Fig 7A). The share was spear-shaped and the mouldboard had little twist. A fairly substantial wheel carriage supported the beam and the drag chain had alternative attachment points on the beam. In 1721, Jethro Tull describes an improved version of the Hertfordshire plough in *The Book of Husbandry*. Then in his *Compleat Body of Husbandry*, Thomas Hale, in 1756, enumerates specific improvements to the same design.

According to Hale the spear-shaped share had been replaced by one broader and flatter, to make a horizontal cut, and the share blended into an iron mouldboard; the handles had been improved in their construction and attachment; the groundwrest was extended to be attached by a 'drock' to beam and handle for more support (Fig 7B).

John FitzHerbert mentions the Kent plough (Plate 10) in the mid-

Plate 10   Kent turnwrest plough (Rutland Museum)

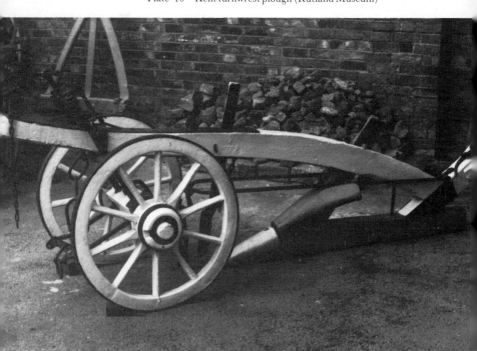

sixteenth century. This was an example of the heavy construction that was successful on the chalk and marsh of south-east England (Fig 7C). The general construction and design was not vastly different from the ploughs of the Midlands and the West Country, except that this was a 'turnwrest' or 'one-way' plough.

The name can be misleading. With an ordinary plough that turns its furrow one way only (usually to the right), it is necessary to arrange a ploughing pattern so that at the end of a furrow the plough is taken to another point for the return cut. In this way groups of furrows can all be turned the same way. If this plough was taken back alongside the furrow just made, it would either turn the new furrow on to it or away from it, depending on which side of the first it was made. Several arrangements have been devised to allow the plough to cut the other side on the return journey and turn the furrow the same way as the previous one.

Where the change of cut is achieved by tilting the share the other way, this is called a 'turnwrest' plough, because the groundwrest is turned the other way with the share, or a 'one-way' plough because it turns all furrows one way—not because it only cuts one way, as might have been expected.

Heavy ploughs of the Kent type continued with little change for 200 years until lighter ploughs were first developed in England, on the east coast. Blith, in the eighteenth century, had a hand in their design and development, and he mentions improving upon the Dutch ploughs, so it is likely the Norfolk, Suffolk and Essex types were based on originals brought over from the Netherlands. In the eighteenth century a trade was building up with the Netherlands which had similar conditions to most of the flat country of East Anglia. A Norfolk plough (Fig 7D) could be pulled by two horses. There was still the high gallows, or fore carriage with large wheels, but the share was more like a modern one and the mouldboard was given a twist to match the curve of the slice, while a groundwrest followed the bottom of the furrow, in a very similar way to a modern plough.

Wooden ploughs were wheeled or swing types, although heavy ploughs needed the large wheeled carriage to support them. An in-between type was the Sussex foot plough (Plate 11), where a sliding foot or skid took the place of wheels, giving the effect of the lighter swing plough, with an adjustable means of giving a forward control of depth (Fig 7E).

Plate 11 Sussex foot plough. The foot is the skid at the front (Jackson Collection)

One specimen of a large wood and iron Guernsey plough (Fig 8) had a large-wheeled carriage to take the pull and, in effect, the plough proper is towed from it in the manner of the Kent and other early ploughs. In this case the power was provided by four oxen, pulling in two pairs (Fig 8A), with a hake giving some horizontal adjustment (Fig 8B). Some of these ploughs had a swinging wooden link to the whippletree (la barre) as an alternative to the hake (Fig 8C). The beam is clamped into a solidly built up block on the carriage (Fig 8D), but the pull is transferred through a chain to a peg with adjustable positions through the beam (Fig 8E). The coulter (Fig 8F) is of conventional form, but the iron mouldboard (Fig 8G) is deeper than usual and the groundwrest (le sabot, or shoe) is comparatively narrow and long (Fig 8H).

The first recorded example of an all-iron plough seems to have been produced by a Suffolk blacksmith, named Brand, in 1766, and in 1807 there is a record of a Hampshire Patent Iron Plough. This had a lever adjustment for the share, allowing the point to be raised or lowered. Around this time a man named Plenty, of Newbury, obtained a patent for an iron frame to which other parts of the plough (Fig 9A) could be bolted.

The plough that set the form that became accepted until the adaption to power was generally known as the 'Rotherham' plough (Fig 9B), which was the subject of a British patent by Stanyforth and Foljambe in 1730. Joseph Foljambe is believed to have built the first plough of this type, but he sold the patent to Disney Stanyforth, who started manufacture in the Yorkshire steel town of Rotherham. The patent was challenged and set aside, as it was ruled to be an improvement and not an invention.

There were variations on the Rotherham plough, but it was always a light swing plough—considerably lighter than the ploughs it replaced, yet able to do better work. Earlier mouldboards had been in many shapes, including nearly flat. The Rotherham mouldboard was given a scientific shape. Although made of wood, it was sheathed with iron (later steel) plates. The main construction was wood, but the design and

Fig 8

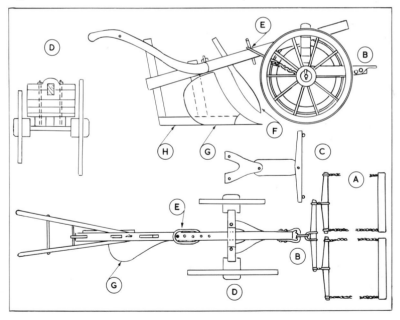

method of assembly gave a structure as strong as the heavy earlier ploughs. Only hake, coulter, share and sole sheathing were metal.

About forty years later a Norfolk farmer, named Arbuthnot, made improvements to the Rotherham plough, without altering it fundamentally, but he does mention the use of the Suffolk hake, which was the first attempt to give vertical and horizontal adjustment at the chain attachment. Arbuthnot also experimented to get the best shape of mouldboard. His method was to use an unsheathed wooden one and let it wear away in use. When it had scoured enough, he made a drawing of it so it could be reproduced in iron.

It was a Scotsman, named James Small, who took a Rotherham plough to Scotland and used mathematical calculations to devise what he considered the perfect shape of mouldboard, which he wrote about in 1784. He was also convinced that much of the strain on the beam would be reduced if the chain pulled from the region of the coulter and was guided only by a form of hake at the fore end. He called this his chain plough.

In America, in 1788, Thomas Jefferson was also concerned about the need to be able to duplicate a good mouldboard and he scientifically produced a mouldboard shape that could be reproduced in wood with saw and adze. Both this and Small's designs came towards the end of the plated-wood mouldboard era. They probably had an effect on quantity-produced iron mouldboards, but they came too late to be of great value to wooden plough building.

The skim coulter is credited to a man named Duckett, from Esher, in 1767, when it won an award from the Royal Society of Arts. While of use in routine ploughing for burying surface rubbish, it also became a valuable extra in match ploughing, where a clean finish was important.

A name still familiar in the manufacture of ploughs, and many other implements, comes into the story when Robert Ransome of Ipswich obtained a patent for tempering cast-iron plough shares. In 1803 he produced the chilled share, making the bottom harder than the top, so that the top wore away and kept a sharp edge. In 1808 he obtained a patent for a plough consisting of parts bolted together, so that a replacement could be fitted for something broken. This idea spread and led to parts for duties other than ploughing being available, to increase the usefulness of the tool.

Influenced by the Rotherham plough, Ransome produced Lincoln-

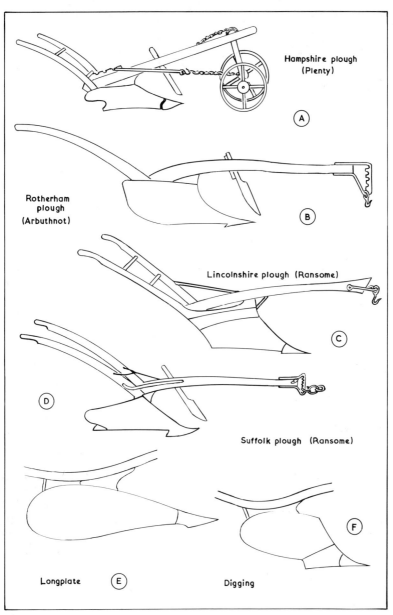

Hampshire plough
(Plenty)

Ⓐ

Rotherham
plough
(Arbuthnot)

Ⓑ

Lincolnshire plough (Ransome)

Ⓒ

Ⓓ

Suffolk plough (Ransome)

Longplate   Ⓔ

Digging

Ⓕ

Fig 9

shire (Fig 9C) and Suffolk (Fig 9D), swing ploughs that bore a family resemblance to it. Ploughs tended to have county names. Ransome's
Rutland plough (Plate 12) was advertised in 1834,

> Each wearing part is contrived that it may be replaced by the
> ploughman without sending the plough off the field. The shares are
> case-hardened on the under side to the temper of steel, which
> occasions their wearing to a thin cutting edge while at work.

Ploughs need different characteristics in autumn and spring work. A
'longplate' plough was used in the autumn (Fig 9E), when the mould-

Plate 12   Ransome's Rutland plough, 1834

58

board turned the slice without crumbling it, to be acted on by the winter weather. For spring ploughing the mouldboard of a 'digging' plough (Fig 9F) was arranged so that as the soil passed up it and was turned over, it was crumbled and pulverised, thus reducing the further work in preparing for sowing. A longplate plough needed wheels to steady it, and these were arranged on the now familiar crossbar, adjustable on the beam. Some digging ploughs are without wheels. The important thing is that the mouldboard turns the slice almost completely over.

Pioneer settlers in America were fighting to make farms from virgin land at the same time as ploughs were developing along more modern

lines in England. They favoured wheel-less ploughs, giving a broad shallow cut, and massive braced coulters to cut through roots and contend with stones. The Pennsylvania plough of the early nineteenth century had the share forged to an iron bar on the landside. The popular Carey plough of that period, widely used in the eastern states, had English and Dutch types of share.

The name that has followed through in America, like Ransome in England, is John Deere. He was a blacksmith who started his career in Vermont in 1825. In the 1830s he followed pioneers who were moving further west. He first settled in Grand Detour, Illinois, and worked as a general blacksmith for the farmers. The cast-iron ploughs then in use were brought from Vermont, where they had been successful on the sandy soil, but proved unsatisfactory on the heavy soil of Illinois. John Deere devised a highly polished steel surface, using a piece of saw blade, and mounted it on a properly shaped, wooden (later wrought-iron) mouldboard, to produce a plough in 1837 that proved to be the answer the pioneer farmers needed for successful farming in the 'new West'.

Improvements were made and John Deere started manufacturing his 'self-polisher' ploughs. He had to import steel from England at considerable expense until in 1846, Jones & Quiggs of Pittsburg rolled the first slab of plough steel and John Deere moved his factory to Moline, Illinois, where the first ploughs made from American steel were produced in 1847.

In 1840, Joel Nourse of Massachusetts had produced his large 'Nourse Eagle' plough, based on Jefferson's principles. This plough had a small wheel which did not normally touch the ground. It was the most popular plough of the time and sales in the mid-nineteenth century were around 25,000 per year.

In the Pacific states there was much work on hillsides, in vineyards and gardens. The popular plough for this work was Robert Knapp's 'side-hill' plough of 1875. This was a conventional plough with a single wheel.

Pioneers, moving west to cultivate the prairies, developed the ploughs of the eastern states into much bigger tools and supported them on frames with large wheels. Such a 'prairie breaker' of the nineteenth century turned a large shallow furrow and laid it upside-down almost unbroken, to kill the grass. This was pulled by at least eight oxen. Later the land was reploughed, or harrowed and planted with wheat or corn.

Variations on the plough

Although a vast array of ploughs were available during the horse-drawn era, nearly all followed the basic pattern of a share and curved mouldboard with a coulter making the vertical cut. Most turned the furrow to the right and made only one cut in each pass across the ground. There were several attempts to do the job in other ways, and extra equipment devised to allow the plough to be used for other operations and so make it more versatile. Two patents for improvements to the Kent plough were granted: in 1860, G. E. Toomer developed the plough in iron, and, in 1861, W. Busby had ideas about a lever action for altering the direction of cut. Neither appears to have had commercial success.

Other designers considered producing one-way ploughs by making the tool double-ended. There is evidence of wooden ploughs made this way towards the end of the eighteenth century, but an iron plough that achieved some success was the subject of a patent by Lowcock in 1843. In Lowcock's plough (Fig 10A) the basic parts are duplicated in each direction, but the more curved section of the mouldboards pivots at the centre so that they swing in a direction to suit the cut. The shares and ground-wrests are arranged so that the whole plough tilts them clear as the direction is reversed. The ring on the end of the draw chain can slide from one hake to the other as the horse changes direction. While the horse is turning, the ploughman swings the handles over to the other direction.

Other designers achieved one-way ploughing by making what amounted to two ploughs, with the one not in use travelling above the other. The idea carried over into steam ploughing, and almost certainly preceded it, although there are no patents as evidence of early types. Horse-drawn balance ploughs are known to have been in use before 1856.

In the balance plough (Fig 10B) the whole operative assembly is duplicated and arranged to be changed over as the handles are swung across. Improvements were the subjects of patents, but these were in the 1870s, when multiple balance ploughs were beginning to be used for steam ploughing, and the improvements were more applicable to them.

Instead of turning endwise, it is possible to change over the operative parts by rotating them on a round beam. If the wheels are not of the same diameter, they also have to be turned on a central rod, or duplicated and turned over (Plate 13). The whole plough is turned and the horse always

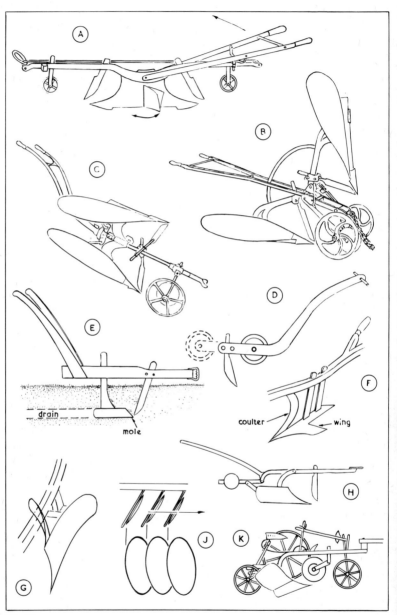

Fig 10

pulls from the same end. To distinguish it from a balance plough, this is called a turnover plough (Fig 10C). A patent was issued to Campbell and White in 1840, in which they used a hind wheel as additional support. Many other patents followed—in fact the turnover plough seems to have been the subject of more patents than any other type—but the forms appear to have stabilised by the beginning of the twentieth century, although there is a Ransome patent dated 1910. There are multiple tractor-mounted ploughs in use today using the same idea.

Thought was given to ploughing more than one furrow at a time with horses, but really successful multiple ploughs had to await the coming of steam and tractor power. Double furrow ploughs were noted by Blith in the seventeenth century. These were heavy wooden types, with shares and coulters staggered at a suitable width, but, except on light soil, this type of wooden plough required a large team of animals to pull it. Iron ploughs using the same idea followed (Plate 14). With the more scientific designs and the lighter construction, such a plough was a more practical proposition, with the ability to do nearly twice as much work in a given time. Three-furrow ploughs were built during the early part of the nineteenth century, but took too much pulling to be worthwhile on most land with only horses to supply power.

Plate 13   Turn-over plough—the parts are duplicated and the machine turns over to plough the other way (Jackson Collection)

Plate 14   Iron two-furrow horse-drawn plough, with parallel beams for width adjustment
(Farmland Museum, Haddenham)

Ploughlike tools were devised for other jobs, but with the coming of iron ploughs, on which the parts were assembled by bolting, it was possible to remove the ploughing parts and replace them with those for other purposes. Several types of shares and mouldboards could be used. Besides the longplate and digging types for opposite ends of the season, a paring plough was used to remove turf and surface growth in the way that was previously done with a breast plough. The peculiarity of a paring plough was that its share cut a broad slice and the mouldboard stood the turf on edge and not right over. When the turf had dried, it was burned and the ashes spread on the land.

The advantages of draining land were appreciated from early days, but drains were laboriously cut by hand. Blith mentions types of draining ploughs used in the seventeenth century (Fig 10D). These did no more than make cuts with a kind of coulter, but they speeded draining by cutting the sides of a trench that could then be dug out.

Several mole ploughs were produced around the beginning of the nineteenth century. These cut underground passages for water to drain through, with a pointed iron cylinder, called a 'mole', and a coulter preceding it to provide clearance for its narrow bar to a beam which served in the same way as that of a plough (Fig 10E). Although there were variations, some the subjects of patents, this was the way they functioned. The power needed to pull them was considerable. For a direct pull there would need to be a large team of horses. Instead, it was more usual to have a form of windlass, operated manually, or by a horse walking in a circle.

At the Royal Show of 1851, John Fowler exhibited a plough which laid wooden pipes threaded on a wire rope as it cut the underground hole. The rope was withdrawn at the completion of the job. All previous mole ploughs had merely left a drain formed of the compacted earth, which might not last very long. Fowler's method has not been improved on, in principle, for pipelaying to this day.

Water can be prevented from draining away by tightly compacted subsoil. Ordinary ploughing turns over the topsoil without disturbing or opening the subsoil. An early way of dealing with the problem was to

Plate 15   Iron ridging plough (Farmland Museum, Haddenham)

follow the ordinary plough with a simple plough that had a deeply set share and no mouldboard. This went along the bottom of the furrow and disturbed the subsoil without bringing it to the surface. This was improved on by John Smith of Deanston who had taken over a marshy farm in 1823. He laid deep parallel drains, but to break up the subsoil in the land between them he made a heavy iron plough, with a share carrying a wing and a curved coulter above (Fig 10F), pulled by six horses. The obvious course of action was to combine the subsoil plough operation with that of ordinary ploughing, but this did not become feasible to a worthwhile extent until the coming of motorisation.

The earliest ploughs threw the earth they lifted both ways. While this was not wanted in later ploughing, it could have an advantage in ridging crops, such as when 'earthing up' potatoes. In the nineteenth century, ridging ploughs with double shares (Fig 10G) came into use for raising the earth around crops which had appeared above the ground (Plate 15).

Both the share and the coulter depend on comparatively fine cutting edges to do their job properly. This means that they are vulnerable when used on land containing stones and tree stumps. The problem has been particularly acute in newly developing countries. In pioneer days in Australia and America, farmers had to contend with land very different from that of the British farmer, who might be following on where ploughs had been used for centuries. Americans used a fairly massive 'prairie breaker' share, which cut quite shallowly and got through the surface to expose any problems below, without too much risk of cutting edges being damaged, as they would be in deep cutting.

Another way of dealing with the risk of hitting rocks and stumps was to cushion the shock to a limited extent. The share and mouldboard assembly of an Australian 'stump-jump' plough, credited to A. B. Smith of Adelaide in 1876, could swing back if it hit an obstruction, and was returned again with a counterweight (Fig 10H).

Disc ploughs represent the only really different approach to breaking up the land. Instead of a share and mouldboard, there is a disc (several on one machine) set at an angle to the direction of pull. The disc is on a sprung arm, so it can lift if it hits an obstruction (Fig 10J). Disc ploughs are more often used with tractors, but they were also horse-drawn, particularly in the pioneer countries. The discs penetrate the surface and turn the soil aside, but they do not turn over a furrow, so they are not as efficient at burying weeds and other surface rubbish.

Up to this point in the development of the plough the ploughman was expected to walk. This meant covering a considerable distance every day on an uneven and usually wet and muddy surface. Attempts to devise ploughs on which the ploughman could ride never met with much success in Britain, but in America, where the areas being ploughed were much more extensive, successful horse-drawn ploughs with seats came into limited use and led up to the types of plough suitable for mechanical power.

In America the first riding ploughs had two wheels and the weight of the rider was taken by a sole that slid alongside the main part of the tool on the unploughed surface. The first of many American patents for riding ploughs was in 1844. Much later, a third wheel was added—two wheels in tandem following the bottom of the furrow, while the other was on the unploughed land (Fig 10K). These machines had a disc coulter. It was discovered that the three-wheel plough could be made more stable and able to plough a straighter furrow if the furrow wheels were inclined towards the landside. Later ploughs made this incline adjustable by the rider who was given lever controls of other functions while travelling.

The three-wheeled plough had become much more of a machine, which was only waiting for mechanical power to make it even more effective. Horses played a part in controlling the depth and performance of the traditional type of plough by the angle of pull. Here they had to provide motive power only, so that the pull was from the end of a beam and the rider did all the controlling with his levers.

Steam ploughing

The most successful early application of steam in farming was to ploughing. Before steam engines were self-propelling, and had to be hauled into position by horses, schemes for using them to haul ploughs across fields by cables had been devised. The idea probably came from the use of a cable and windlass to haul a mole plough. In 1810 Major Pratt obtained a patent for steam hauling a plough. John Heathcoat, in 1836, spent a very large sum in developing a cable-hauled plough, with the engine on giant crawler tracks, intended for working and reclaiming marshland, but this was never commercially successful.

It was John Fowler's ideas and Ransome's production methods that had most effect on the development of steam ploughing by cable. Fowler showed his mole plough drawn by a steam engine at the Royal Agricultural Society Show at Lincoln in 1854, and Fowler and Ransome produced steam ploughing outfits in 1856.

The plough used for Fowler's, as well as nearly all other systems, was a balance type. Early ones would cut three furrows at each pass. As power and efficiency increased the number of furrows went up. Six at a time was common and as many as eight became possible with larger engines.

The balance plough (Fig 11A) was carried by two large central wheels. The plough parts were duplicated at each side and mounted at first on wooden beams, but later on steel girder assemblies with a small land wheel near each end. The two halves were angled upwards, so the side out of use swung clear of the ground. Each ploughing assembly consisted of the usual share, mouldboard and coulter. The assembly at each half was set at an angle to the side, so the shares cut furrows at a suitable distance apart. In some ploughs the parts could be moved or removed and, with the amount of power available, subsoiling attachments could be included. A man travelled on the plough and steered with a wheel and worm gear. His seat and equipment were duplicated towards each end, where his weight helped to keep the working end of the plough down.

One snag with a balance plough that was exactly on balance was that the part in the air tended to bounce and sometimes prevented the working end from digging as deeply as it should. This was overcome in the Fowler anti-balance plough in 1885, when the central assembly on which the whole plough pivotted was pulled by the cable to an off-centre position at each reversal of direction. This ensured the part on ground being sufficiently heavier than the part in the air.

For cable ploughing with a single engine there had to be an anchor at the opposite side of the field (Fig 11B). The anchor was a heavy wooden carriage that could be further weighted with earth, running on four sharp-edge wheels that pressed into the ground and resisted the pull on the drum taking the cable (Fig 11C). At first this had to be moved by hand, but Fowler devised a self-adjusting anchor. After each bout, both engine and anchor had to be moved the correct distance for the next set of furrows. Of course, the plough was of the one-way type that turned all furrows the same way.

The use of two engines was more satisfactory. One engine hauled the plough across the field while the other paid out a slack cable, and at the same time moved on along the headland for a sufficient distance to be

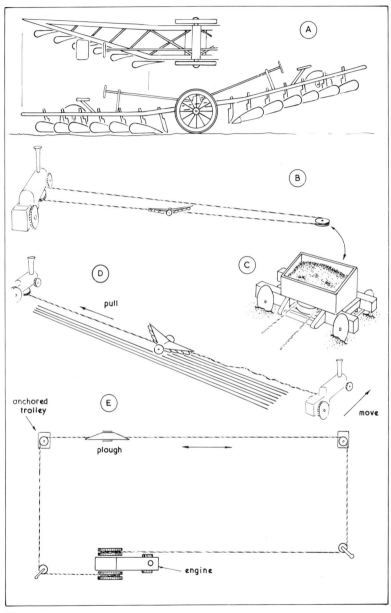

Fig 11

ready to haul the plough back to make the next line of furrows (Fig 11D). This system was quicker to set up and the two engines could carry all the equipment between jobs, so it was favoured by contractors, despite the greater initial expense.

Another method allowed the engine to remain in one place, while the plough was drawn backwards and forwards between two anchors (Fig 11E). These had to be moved progressively down the field and there had to be various guides for the moving cable, so rather more labour was needed in setting up and servicing the rig. Early rigs used fibre ropes, which were troublesome and flexible steel ropes later made the work much more efficient.

Tractor ploughs

As steam engines became lighter, due mainly to the use of high pressure boilers, it became feasible to use them as tractors for pulling ploughs and other implements. This stage was reached not far in advance of the coming of the internal-combustion-engined tractor, so that implements and attachments developing for use with steam tractors continued to be improved and perfected for use with the new method of power farming.

Horse ploughs, particularly of the two-furrow type, were adapted to be pulled by tractor, while the three-wheel riding type, more popular in the USA, showed the way development was going, with lever and other adjustments so that faults in performance could be corrected by the rider without stopping the machine. Early towed ploughs required a second man on them as well as the driver of the tractor.

Later tractors had the controls within reach of the tractor driver. Such a two-furrow tractor-drawn plough coupled to the tractor hitch with a clevis (Fig 12A). Two wheels provided the main support for an iron framework carrying two shares and mouldboards (Fig 12B), attached by frogs (Fig 12C) with disc coulters (Fig 12D) on adjustable supports. The whole frame could be raised or lowered by a lever (Fig 12E), or tilted by a handle (Fig 12F). The rear wheel (Fig 12G) was tilted so it ran against the side of the furrow and took some of the strain due to the ploughing action. This is even more necessary because many of these ploughs are without landsides behind the mouldboards.

Plough attachments coupled directly to the tractor are now more

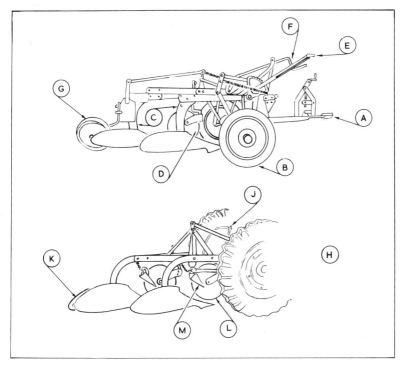

Fig 12

usual, and these came into use with steam tractors. Today a large number of furrows may be cut at each pass, but with lower-powered tractors the two-furrow plough was common (Fig 12H). The plough is matched to the tractor by the three-point system pioneered by Ferguson, which is designed to include provision for hydraulic adjustment of lift and depth control. In the example a lever provides sideways adjustment (Fig 12J). There is a sloping wheel (Fig 12K) behind the rear mouldboard. The disc coulters (Fig 12L) are fitted with skim coulters (Fig 12M), which serve the same purpose as on the horse plough, cutting off surface weeds and grass ahead of the main cut.

Such a directly mounted plough can be simpler in construction. The whole outfit is shorter and there is the weight and steadiness of the tractor contributing to maintaining a regular depth of cut. With the shorter outfit each batch of furrows can be taken nearer the edge of a field and with the plough lifted the tractor turns in a small space. The resulting narrow headland is easy for the plough to be run around after completion of straight ploughing.

Four-furrow directly mounted ploughs are usual today, although other numbers are possible. Furrow widths are 12in, or slightly more, and can be adjusted by the driver. For four furrows the tractor has to be about 70hp, while at least 80hp is needed for five furrows. A modern fixed multi-furrow plough has a substantial, but uncomplicated, backbone and there is provision for securely fitting, different bases, consisting of mouldboard and share, to suit the type of land, the work to be done and the speed of working. Coulters can also be changed, and may be knife or disc. Skimmers can also be changed. A control wheel running on top of the unploughed landside may have a screw adjustment to give control at the remote end of the backbone.

Although directly-mounted multiple ploughs are widely used, they

Plate 16   A Ransome's TSR 108 modern multiple turn-over plough

offer the same problem as the basic horse plough in having to be worked to a pattern on the field to get groups of furrows turned the same way. Some variations based on the one-way horse ploughs are used on tractor ploughs. Popular are the turn-over multiple ploughs (Plate 16), which can be rotated so that the plough turns furrows to the other side and the tractor can return alongside the work just done. These are now more logically named reversible ploughs.

A reversible plough is similar to a fixed plough in having a backbone able to carry bases, coulters and skimmers which can be changed, but all of this is duplicated above the backbone, so when the whole thing is turned over by power, the second set of equipment arranged to cut the other way comes into use.

CHAPTER 5

Planting and Sowing

Early man must have scattered his seed broadcast by hand on his scratched ground, then scraped earth over it with a bush.

Seed was carried in a cloth fastened to the waist and shoulder and held by one hand, while the other hand scattered the seed. This was followed by a basket, called a hopper or seed-lip. The basket was held to the waist, suspended by a strap. A handle at the forward side was used by one hand to steady it. Some baskets were woven in the typical over and under method around upright pieces. Others were made by coiling strips of hazel, willow or other pliable wood and sewing over them in a method still called lipwork, which may have come from this early use. Other seed-lips were made of thin wood (Fig 13A), and the better seed-lips were kidney-shaped so as to fit more comfortably against the body.

The ordinary seed-lip was steadied by one hand while the other was used for sowing. Some workers favoured a seed-lip held tightly by straps in front of them so that they could sow alternately with each hand. Whatever method was used, the sower had to adopt a steady sequence of step and hand work to get an even spread.

An aid to even sowing was the use of a mechanical device, generally known as a fiddle because its method of operation was like drawing a bow over a violin. It was no quicker, but it reduced the amount of skill needed. A seed fiddle, still occasionally used, had a box, extended in its capacity by a bag, possibly strapped to the shoulder. The device was slung by a strap and operation of the handle spun a star-shaped wheel in alternate directions. This threw seed in a regular pattern. The rate at which the seed dropped on the wheel could be adjusted by a lever (Fig 13B).

Holes made by dibbers were used for garden crops, but they were con-

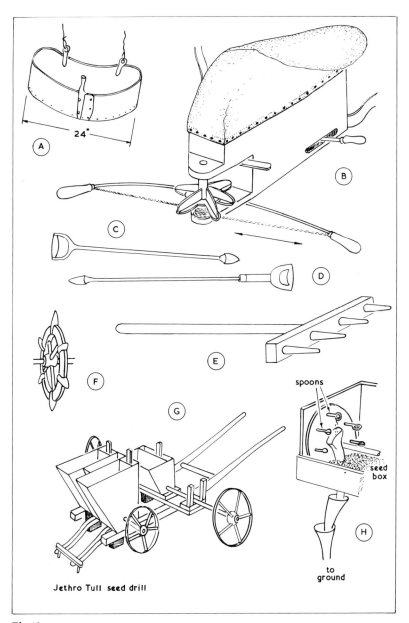

24"

A

B

C

D

E

F

G

spoons

seed
box

to
ground

H

Jethro Tull seed drill

Fig 13

Plate 17 The wheels of a potato planter which are adjustable on the axle to control row spacing, while the 'dobbers' may have their position adjusted around the slots in the wheels (Farmland Museum, Haddenham)

sidered as an alternative to broadcasting seed in the field until early in the seventeenth century. A book called *A new instruction of ploughing and setting of corn* by Edward Maxey in 1601 describes planting corn with the use of a dibber. His method was surprisingly advanced technically, if excessively tedious. It involved using a board with holes at the correct spacing, with a dibber to push through the holes and be stopped at the correct depth by a shoulder. Seeds were dropped in, the board moved on along a stretched line, and earth was raked over. The more usual method that became accepted was for one man to walk backwards using two dibbers while another followed, dropping in seed and covering over.

Dibbing was normally kept to reasonably straight lines and this made hoeing between the growing corn simpler. Corn dibbers may have been wood, but most that have survived were made by the blacksmith, with an end up to 3in (7cm) long and 2in (5cm) diameter on a thinner stem taken to a handle at a convenient height (Fig 13C). The handle may have been iron, but a wooden one was more comfortable to use (Fig 13D).

Attempts to secure even spacing without the complications of Maxey's board were made. There were double dibbers, held at the intended distance with crossbars. The crossbar had the advantage of allowing the foot to be used as a relief from hand pressure. A variation had several dibber points along a board, which had a handle for lifting and locating. Pushing this by hand or foot produced a series of holes (Fig 13E). A further step was the use of a hand or horse-drawn roller, or wheels fitted with spikes to make the holes (Fig 13F). The more ingenious models had the spikes adjustable to give desired spacings along the rows, and there were many wheels adjustable to make several rows at the right spacing (Plate 17).

Two actions, usually by two workers, were needed to plant with a dibber. Attempts were made to bring the two actions together by dropping seed through a hollow dibber, but this was complicated by soil and mud choking the end. One type that had some success carried the seed in a hollow dibber. When the user released a catch the end opened and let out a predetermined number of seeds. However, by this time mechanical planting with a seed drill was coming in and dibbling, whether improved or not, had had its day.

Making grooves or furrows in the field was practised with broadcast sowing. This was done with a ridged roller, or by ploughing with an ordinary or ridging plough. It was an aid to getting at least some of the seed to a reasonable depth, where the soil was drawn over with a harrow, rather than an attempt to plant the seed in rows. There had been some examples of devices that allowed seed to be dropped in as a plough opened a furrow. A Babylonian drawing of 1316 BC shows an ox-drawn plough with a funnel, or hopper device, into which seeds are being dropped as the plough progresses. Other earlier civilizations certainly had devices using similar ideas, but in British agriculture, devices, which became known as seed drills, did not come into use until at least the sixteenth century.

It was Jethro Tull, with his book *New Horse Houghing Husbandry*, published in 1733, who is credited with the introduction of practicable seed drills, yet this was only regarded as secondary by him. He was more concerned with getting seeds into rows so that horse hoeing could be used. To him the method of planting was only incidental. Tull, a lawyer, experimented with seed sown by hand in drills to discover the best depth.

77

From his scientific approach to the problem of sowing he devised both the hoe and the drill, but he met with opposition and ridicule, and his ideas were not fully appreciated until after his death in 1741.

Tull's first seed drill made channels for three rows of seed. The seed from a hopper was fed to a grooved cylinder rotating against a spring-held tongue. From the cylinder the seed was fed to tubes leading to the soil. The tongue had a likeness to a part of an organ mechanism. The drill could be pulled by one horse. Power for Tull's drill came from the ground wheels, of which there were four. The two larger front wheels took the weight of the machine and carried a hopper, while also working the front drill. Two smaller wheels worked the other two drills (Fig 13G).

Later developments by other experimenters used what became known as either force feed or spoon (cup) feed in the vital mechanism, which metered the amount of seed fed to the planting tubes. Both methods had been anticipated in less-successful machines produced before Jethro Tull's invention.

Plate 18   A mangold drill, showing the funnel-shaped feed pipes leading from the metering device in the hopper

In a seed drill, as it developed to the present day, the internal seed hopper feeds seed via a metering device to tubes leading down to the soil, where a coulter makes a furrow just ahead of the seed dropping (Plate 18). The soil crumbles over the seed, but covering is usually completed by harrowing. The number of rows that can be sown at one time is variable and with tractor power may now be quite large. The distance between the rows can be adjusted. For grain it may be only a few inches; root crops and vegetables could be up to 2ft (60cm).

In an early version of the cup feed system the seed from the hopper kept a seed box topped up. Rotating discs carried tiny spoons or cups through the seed in the box. At the top of the turn seed fell from the spoon into a funnel leading to the coulter tube and the ground (Fig 13H). In the force feed system the seed is taken from an opening at the bottom of the seed box, which can be adjusted, and pushed towards the outlet to the coulter tube by a rotating device. Several arrangements are used. One is a roller with spaces between projections to take the seed (Fig 14A).

Many seed drills were devised during the century following Tull's machine. Drills that broadcast seed instead of putting it in orderly rows were devised. This has advantages for grass and clover and is still practised. Steerage seed drills, working in the same way as steerage hoes, were made so that a man walking behind the machine could keep the drills straight when the horse deviated slightly. Some seed is better and more economically planted at intervals and machines were made that automatically interrupted the flow to get this result.

Modern tractor-drawn seed drills may be more sophisticated and better engineered machines, capable of covering a broad band at one time, but they function in the same way as machines of the horse-drawn era. One feature is the use of a marker disc. This disc coulter travels at the side of the machine and makes a small furrow (Fig 14B) at a distance so that the driver knows if he keeps the tractor wheel along this line the drill following will match up with the piece previously worked.

Potatoes needed different treatment and for a long time after seed planting was mechanised they continued to be sown by hand. A seed potato (tuber) that has sprouted (chitted) needs more careful handling than can be expected from a machine, although when planted non-chitted there is less of a problem. Machines have developed from the latter part of the nineteenth century, but even modern potato-planting machines are not fully automated. In one type (Fig 14C) operators ride

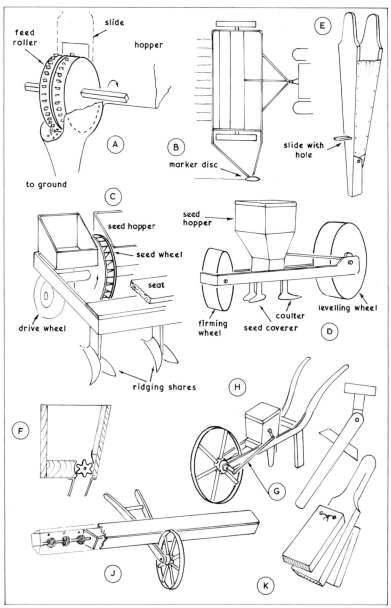

feed roller

slide

hopper

A

to ground

B

marker disc

E

slide with hole

C

seed hopper

seed wheel

seat

drive wheel

ridging shares

seed hopper

firming wheel

seed coverer

coulter

levelling wheel

D

F

H

G

J

K

Fig 14

on the planter and place individual sprouted potatoes in compartments in a shielded wheel. A small share makes a furrow ahead of each wheel, which deposits potatoes quite close to the ground so as to minimize damage to the shoots. Ridging plough shares or offset discs follow to cover the potatoes.

In a machine that deals with non-chitted potatoes, they feed from a hopper on to compartments on a moving belt that serves a paddle conveyor which drops potatoes over the furrows through holes. The furrow-opening shares and ridging plough shares are the same as on the other machine. On this near-automatic machine an operator rides and himself puts potatoes that have been missed into conveyor cups, or takes potatoes out where two small ones have gone into one cup.

Sugar beet is a more specialized crop that once involved a considerable amount of labour. In recent years there has been much mechanization.

Plate 19   Jabez Buckingham root drill. The seed-dropping mechanism is driven from a ground wheel (Jackson Collection)

Typical is the continuous-flow seed unit (Fig 14D). Several of these are towed at the correct spacing from the tool bar of a tractor. Each unit has a large levelling wheel. Behind this is a hopper with an agitator, or other metering arrangement, pushing seed through a hole, which can be adjusted by turning a disc. Below this a pair of shares open a furrow before the seed drops and another pair close the furrow. The trailing wheel firms down the soil. Drive for the metering device comes from a shaft worked through bevel gears from the big levelling wheel. More advanced units have greater precision and can space seeds accurately, either individually or in small groups. As well as sugar beet such machines are used for turnip, cabbage and other vegetable seeds.

Besides the horse-drawn seed drills, which developed into the tractor-drawn drills, there were smaller hand-operated arrangements to give better seed distribution than freehand planting. Some of these were carried. In one the seed dropped through a hole in a plate and out through the bottom of the planter as handles were moved with a bellows action (Fig 14E). Other planters were wheeled and dealt with a single row at a time (Plate 19). Improved versions have survived to the present day for market garden use, particularly for such seeds as peas and beans.

The many hand-wheeled planters varied, but basically a single wheel drove a metering arrangement in a hopper, either a rotating fluted wheel (Fig 14F), or a reciprocating flap valve (Fig 14G). The wheel may have made a sufficient furrow or there would be a coulter ahead of the seed tube, and the whole thing had legs like a wheelbarrow (Fig 14H).

The same idea was used until quite recently for the broadcast sowing of grass and clover. The barrow carried a very wide (8ft or more) trough for the seed, with many small brushes rotating on a long axle (Fig 14J). These threw the seed against holes in the rear of the trough, to spray fairly evenly as the barrow was pushed. Some arrangement was provided to regulate the flow of seed by adjusting the sizes of the holes.

Seed drills were combined with other machines. Some towed a harrow to cover the seed. Planters were made to mount on ordinary ploughs, with their own wheel to drive them and feed seed into the furrow as it was made. Some seed drills had demountable parts so they could be converted to hoes later in the season.

A problem associated with sowing seed is the prevention of too much of the seed or young sprouting plants being eaten by birds. Scarecrows go back in use a long way. Various devices to produce noise have always been made, eg boys were employed to shout or otherwise make a noise. Common noise-makers in the last few centuries were clappers (Fig 14K), with two boards loosely pivoted by cords on each side of another, or a swinging clapper. More recent devices use explosive caps or cartridges.

Fertilizing and spraying

The distribution and application of manure was by cart and hand tools, with little attempt at mechanization evident until well into the nineteenth century. As the value of liquid manure became appreciated, stock yards were arranged so that rain and household waste water also flowed through the yard to filters and a holding tank, from which the liquid could be pumped to the cart. A cask of something like 150 gallons

Fig 15

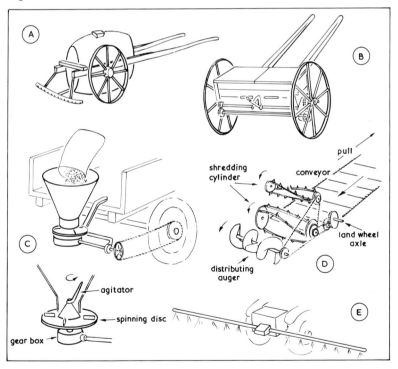

capacity was supported between shafts (Fig 15A) and the axle for the large wheels was shaped below it. Filling was by a funnel built over an enlarged bung-hole. Distribution was via a perforated tube extending the full width of the cart. Metal tanks mounted on tractors were later used for the same purpose.

Dry powdered manures, such as bone dust, lime, soot and compounded fertilizers were sometimes distributed by hand from a seed-lip, but even distribution was difficult if there was a wind and could be uncomfortable or even hazardous to the health of the worker. A machine was invented by Alexander Main in 1839, with a fluted roller turned by the land wheels onto which the fertilizer dropped and was limited by a rotating brush. The fluted cylinder fitted closely to a chute and the powder was dropped onto the land through shields taken low down to prevent it blowing about. Other inventors produced variations on this. A later version, of the same general form, has a large wooden hopper, inside which mechanism, driven through gearing from the land wheels, stirs and forces the powder out through slots. The rate of distribution is controlled by a lever (Fig 15B).

A more successful idea used one or more spinning horizontal discs. In such a centrifugal manure distributor (Fig 15C) there is a cone-shaped hopper, which may be on the back of a trailer which carries sacks of powdered fertilizer. When a shutter is opened at the bottom of the hopper the powder falls through on to a spinning disc. This may be driven through chain and gearing from the land wheels or from the power take-off of a tractor. An agitator may turn inside the hopper to keep the powder free-flowing and an adjustment at its base controls the amount of powder. The speed of rotation of the disc affects distribution—a higher speed spreading the powder wider and thinner.

Spinning discs are used today, usually directly mounted on the back of a tractor and with a hopper large enough for the chore of tipping sacks to be dispensed with. Such a device can be taken at about 5mph and will spread evenly over a width of maybe 28ft (8m). The same tool can be used to broadcast grass and cereal seeds, becoming then a mechanized, sophisticated, larger version of the fiddle (Fig 13B, p 75).

Mechanical distribution of solid farmyard manure did not become successful until tractor power became available. The process is now usually called muck spreading. The equipment consists of a means of carrying a large quantity of dung and machinery to scatter it in small fragments as the machine progresses. In one type a near-cylindrical tank, in line with the towing tractor, throws particles to one side. The more common type in general use throws them in a broadcast fashion to the rear.

In this type of muck spreader the bulk is carried in a long trailer, with a spreader floor that is a conveyor belt moving slowly towards the back of the machine. At the rear the manure passes against shredding cylinders which tear it apart and throw it against a rotating distributing auger (Fig 15D). The effect is to send out a shower of small manure particles in a fairly even distribution to the rear of the machine as it is towed along. Power may come from a land wheel or from the power take-off of the tractor. The conveyor belt floor moves in a jerking manner through a ratchet and cam drive. This can be adjusted to vary the rate of application of the manure. Both water and liquid manure may also be spread over a considerable width from the back of a tractor (Fig 15E).

CHAPTER 6

Cultivating

When primitive man turned from collecting the products of nature as he found them to attempting to grow crops under his control, he must have started preparing the ground by scratching with a stick, then realized that if he was to get the most successful crops he would have to break up the earth to a greater depth. In the early stages he would have dug, with his stick flattened to a wedge shape with an axe, or with a naturally shaped spadelike end.

Spades

A wooden-bladed tool cannot be expected to stand up to digging in even the finest soil for long without breaking or wearing away, so with the coming of iron, wooden spades were either tipped or completely sheathed in iron (Fig 16A). Wood without metal sheathing only persisted as shovels for grain and similar things, where the edge did not have to cut its way into the mass.

The next step was to make the blade entirely of iron. With minimum facilities for working iron this meant forming a handle socket by wrapping the metal around the wood (Fig 16B). Sheet metal lacked stiffness, so a turned-up edge gave both stiffness and a square edge to the cut (Fig 16C). If the foot was to be used on the tool, it was put on a peg (Fig 16D), a notch in the handle (Fig 16E), or a turned-over edge on the blade (Fig 16F). Spades for cutting peat today follow these patterns.

Spades were first made by local blacksmiths to suit what local workers regarded as best for their own soil. Manufacture eventually settled in a few industrial areas, but these producers of tools in quantity were unable to convince their customers that a spade used in one part of the country

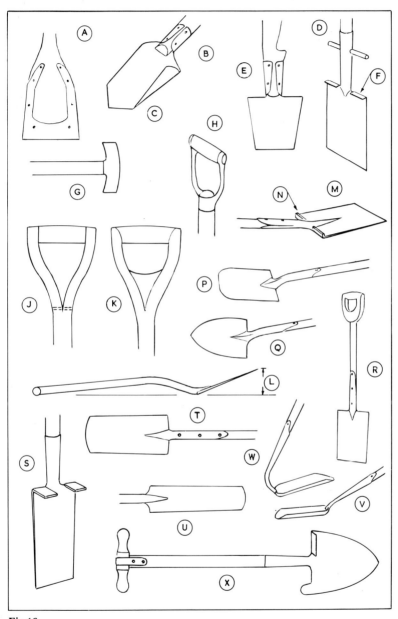

Fig 16

should be just as successful in another. Variations in pattern, to suit local traditions and preferences, still had to be produced. Even today catalogues show regional types of spades and other tools.

Spade handles would have been mostly poles just as they were cut. Of the British woods, ash is most suitable for taking the strains that come on spades and other tools used for levering or for a hitting action, and fortunately, ash can usually be found growing straight enough. Straight poles as handles suit light soils. The British worker has always favoured a shorter haft with some sort of grip at the top, but this is not general and in some countries the straight long handle is used in almost all situations. In France and Ireland straight-handled spades are more common.

The T-handle, mortise-and-tenonned on (Fig 16G), as commonly seen today, was beyond the skill of most country workers, so it may have been held by a metal strap. Alternatively, the blacksmith made a metal bracket to form a D-handle (Fig 16H). Later, manufactured spades had the ash cut and steamed to shape (Fig 16J), preferably with the angle filled (Fig 16K) to strengthen it.

Not all blacksmiths could make sockets, so a spade might have been spiked into the end of the wooden shaft, which was prevented from splitting by a ring or ferrule. This was not very satisfactory and the socket developed from earlier attempts at wrapping sheet metal around.

The basic spade has a rectangular blade with the socket and handle cranked at an angle to it. Individual workers favoured different 'rises' or 'lifts' (Fig 16L). A digging spade is slightly curved in its cross-section to provide stiffness. If it has a name, this is a London spade (Fig 16M). If the top is thickened, it is 'treaded' (Fig 16N). The worker would lace a spade iron below his boot to protect the leather. Yorkshire and Lancashire also laid claim to slight variations on this plain spade, but an Ormskirk spade has a curved end (Fig 16P), while a Southport spade is pointed (Fig 16Q), and a Norfolk spade is tapered (Fig 16R).

There are several versions of long tapered spades. A flat one cuts the sides of post holes (Fig 16S). A grafting tool (Fig 16T) is only moderately longer and narrower than a standard spade, and it may be parallel or tapered. Longer curved versions are draining tools or trenching spades (Fig 16U). A lighter version is more of a scoop for removing loose soil and may be arranged to push (Fig 16V) or pull (Fig 16W). A rutter (Fig 16X) was a very heavy drain-cutting spade, which may still be used in forest draining.

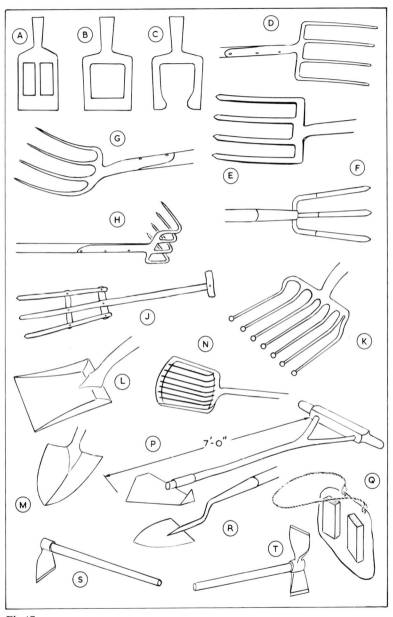

Fig 17

Forks

Forks, as we know them, for digging do not go back very far, probably due to the difficulty of making them strong enough from the steel available and the equipment that could be used. An early semi-fork was a spade with holes in it (Fig 17A). One used for digging clay, called a mule (Fig 17B), developed into a fork (Fig 17C). Most modern British digging forks have four prongs, either of comparatively slim section for general work (Fig 17D) or broader for digging potatoes (Fig 17E). Most early forks were three-pronged, presumably for ease of construction. The Ormskirk area favoured a three-pronged potato fork (Fig 17F) and the Irish seemed to prefer more curve to the top (Fig 17G). A manure fork tended to be bigger than a digging fork and some had more than four prongs.

Bent forks were used as rakes to drag dung off a cart (Fig 17H). Various sizes of these sometimes served as rakes for preparing seed beds or as hoes for earthing up potatoes and other plants. On the Isle of Wight a bent fork was a 'grapple'. In the Midlands it was a 'scratter' or 'Canterbury hoe'. In hop fields it was a 'spike hoe'. Another version, sometimes two-pronged, was used for dragging seaweed to be used as manure. Wooden versions of manure or dung forks have survived (Fig 17J). A multi-pronged fork with ball points was used for shovelling beet (Fig 17K).

A shovel differs from a spade in being a scoop on a handle, for picking up rather than digging. Again, there were many local preferences. Basically, if the end was square the sides were turned up (Fig 17L). If it was pointed, stiffness came from a fairly deeply curved cross-section (Fig 17M). A cross between a shovel and a fork was a 'potato scoven' (Fig 17N) that sifted off some of the dirt.

Earth breaking

What to do with the tangle of grass and weeds on the surface was a problem before the days of ploughs that turned over enough soil to bury it. A common treatment was to take off a thin top layer. The turf could then be burned and the ash returned to the soil. The tool for this job was a breast plough, or 'flaughter spade', which was fairly massive and often crudely made (Fig 17P). The point aided penetration and the turned edge stiffened the blade and squared the cut. Breast ploughing was obviously a very arduous occupation and some relief was given by a leather apron with strips of wood over the chest (Fig 17Q), while the thighs were padded for a lower push. A more refined lighter modern version is called a 'turfing iron' (Fig 17R).

On heavy soil, in particular, it was necessary to follow digging with breaking up the larger clods of earth. This might be done with a mattock (Fig 17S). A double-sided implement was called a two-bill or American mattock (Fig 17T). Later mattocks were sharpened like axes and have changed their purpose to cutting through tree roots. Clods were also broken with a 'beetle', simply a mallet made of a section of log on a handle. The same beetle was used with a wedge for splitting logs. Clods were also broken with a 'clod knocker', a sort of open, rounded spade on a long handle. A similar tool with serrated edges was used to break down nettles. The pick axe has not had much agricultural use. It is associated more with mining, quarrying and roadmaking, but there is evidence that primitive farmers lashed a deer antler across the end of a handle to use as a pick for breaking up earth. Drawings found in Egypt show tools used with a pick action (Fig 18A).

With tools made to suit local ideas, and there being little contact between people only a few miles apart, digging tools varied considerably and might be traced in almost any form. A sort of horned spade with upturned cutting edges was used after the manner of a breast plough for digging drains (Fig 18B). This must have been very tiring work as the drain got deeper. Other spades would have been needed to trim sides and a very stoutly constructed fork was used for breaking up the subsoil for the horned spade to clear out (Fig 18C). Picks were also used, both of the swinging type and with a swordlike blade having a foot tread and a wooden cross handle. Variations on these tools were used in medieval times and up to the coming of mechanical aids to draining.

The Portuguese are credited with going halfway towards a fork in their spade design (Fig 18D). This was claimed to be better on stony soil than a straight-edged spade. Spades and forks were made with their handles arranged bent back or forward for particular purposes, like keeping the worker's hands clear when cutting the edge of a ditch near a wall. It is only in comparatively recent times that the pressures of quantity-production have caused manufacturers to discourage the demand for local and special types of spades.

The halfmoon type of cutting spade now used by gardeners to trim

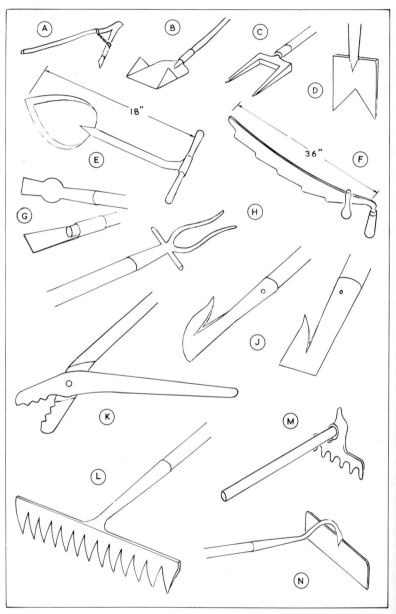

Fig 18

lawn edges had its origin as a turf cutter and was also used for cutting out ants' nests.

Pointed spades, sharpened on the edges, were used for cutting turf and peat, but similar spadelike tools with shorter hafts and broader handles were used for cutting out slabs of compacted manure when it was to be removed from the stock yard to the fields (Fig 18E). The same, or similar, tools were used for cutting hay from the stack. A tool of the same form, but about half this size, was a barking iron for stripping oak bark for use in tanning leather. The large hay knife was also used for cutting manure and the biggest serrated-edged and two-handled type (Fig 18F) was particularly favoured for manure.

While mattocks were used for breaking up large clods of earth, lighter hoes were used to make the soil even finer and to combat weeds, as done by gardeners today. Basically, designs have been refined rather than altered. Early pushing Dutch hoes were often pointed, after the manner of a breast plough, to cut through weeds. Draw hoes were straight across or pointed and were sometimes made with a double point like a Portuguese spade. Hoes with two or more heads on one handle were an early attempt to speed production.

Until a century ago weeds had to be dealt with by hand, although several tools were devised for dealing with individual weeds. Some of these, called 'weed spuds' or 'paddles', were like slim Dutch hoes (Fig 18G) or two-pronged forks to get around a dock or tap-rooted weed and lift it, possibly with the aid of the foot on a tread (Fig 18H). A fairly general tool was a weeding hook (Fig 18J) that cut on the pull stroke. It could also have a squared sharp end for cutting on the push stroke. A farm worker might have one of these on the end of the staff he used when walking, so that he could deal with weeds in passing. Pincerlike tools, either entirely iron or with wooden handles, were used for pulling individual weeds—in particular thistles (Fig 18K)—that would grow again if only cut off.

Rakes

Rakes, as known by the modern gardener, were not used for cultivating soil. Raking a field, from quite early days, was better done with a harrow. A broad rake with tapered spaces was used to drag over grass (Fig 18L). This allowed grass to pass through, but flowering weeds had their heads pulled off. The wood or soft iron of early ploughs suffered from stones and a hand rake was used to gather them ahead of the plough (Fig 18M). This was a job for children, and the stones might then be used for roadwork. A sort of short-handled hoe was carried by the ploughman to scrape away mud from his plough.

Not strictly a rake, although sometimes so called, probably after the similar blacksmith's tool of the same name, was a tool used for pulling dung out of a cart. It was also called a 'harle' or dung scraper (Fig 18N). It had its uses for scraping mud in the farmyard, and in fairly recent catalogues was advertised as a road scraper.

Hoes

Mattocks in various forms, with their family likeness, probably preceded hoes, but in lighter construction. Both tools owe something to the pick type of tool, which was either lashed across, or spiked through, the end of a handle. Until quite recent times manufacturers were producing hoes in many shapes, with spikes that could be used as tangs fixed into the ends of the handles for a pushing action (Fig 19A), or across, through holes, for a chopping action (Fig 19B). In both cases the tools were used for breaking up the surface soil and removing weeds between crops. A variation used a wrapped socket for a handle (Fig 19C), which was easier for a blacksmith to make than a cast or forged socket, as found in more recent hoes (Fig 19D).

The better hoes were given a tapered socket (Fig 19E) to resist any tendency of the head to fly off. This followed the practice used in adzes, mallets and other craftsmen's tools. Narrow hoes were called grubbing hoes, the tapered type being ox-tongue hoes (Fig 19F). Broad hoes, as used by modern gardeners, did not have much use in earlier days and the modern form of Dutch hoe (Fig 19G) does not appear to go back very far.

Strong narrow hoes, similar to the grubbing hoe, were once made with variation of form to suit local preferences and are still so made for use in Africa. The narrow hoe is still used in the Scottish Highlands, where it is called a 'chappie'. In Kerry, Ireland, a similar tool is called a 'graffan'. Both this and the Africa versions are used on long handles in the same way as a pick to break up soil, as an alternative to digging with a spade or a cas chrom.

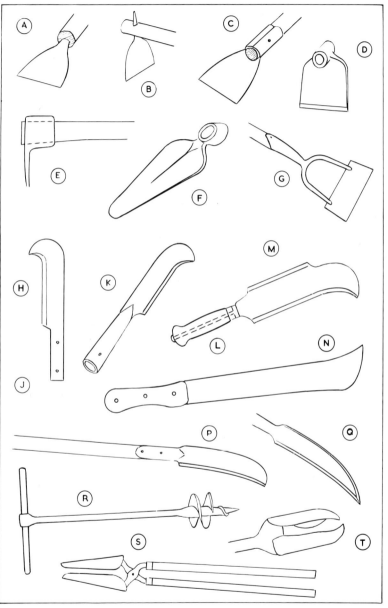

Fig 19

Billhooks

Allied to the tools used directly for preparing and cultivating the ground were those used to clear the ground and prepare and maintain hedges. The axe in its many forms has been used from the earliest days, but as iron and steel became more generally available, workers supplemented their axe with various forms of swinging knives.

The general-purpose type was, and still is, the billhook. Basically this is a knife with a hooked end (Fig 19H). The handle could be a flat tang to take wood on each side (Fig 19J), a socket into which a tapered handle fitted (Fig 19K), or a long tang to pass through a turned handle (Fig 19L), which is usual today. On this basic pattern there have been many local variations of length, curve and size, while a few billhooks have a straight cutting edge on the back (Fig 19M).

On a long shaft the billhook became a weapon. The fighting bill followed the hooked-knife form, but also had a spike pointing forward and another sideways, opposite to the cutting edge. Many of these found their way into agriculture after the British Civil War. The billhook-and-slasher form was used in many specialized tools. A Kent 'hop hook' was a curved billhook with a spike on the back, something like a fighting bill, with the spike used for pushing. Large blades like slashers were used for clearing water courses. Tools of much the same appearance were used by thatchers and for cutting hay or manure.

European farm workers have never adopted the machete (Fig 19N), which serves a very similar purpose to the billhook in tropical countries. These are about twice as long (20in, 50cm) and are used on cane and similar undergrowth, although they have never had the hooked cutting edge of the billhook, which might have been expected to give the same advantages on these tools.

The billhook form on a long handle was a slasher (Fig 19P). The amount of curve to the hooked end varied, but this had an obvious advantage in retaining the edge on the wood when cutting a branch too high to be held. Some slashers were sharpened on both sides (Fig 19Q), and it was usual to have a long socket into which the handle was riveted.

Hole boring

Grown hedges or loose stone walls were the usual means of making boundaries to fields, but where wooden fence posts had to be erected man exercised his ingenuity in devising tools to make holes.

A post-hole borer worked on the screw principle: turned by a cross handle it would penetrate and remove soil (Fig 19R), providing it was free from rocks or large stones. Augers using the same idea are now mounted on and driven by tractors. There were several types devised that operated two opposing scoops with a pincer action (Plate 20) and this two-handled type survives today (Fig 19S). Tight soil may have to be loosened with a pick or spike, but the tool will clean out the hole. Another

Plate 20   A pincer-type post-hole borer

arrangement had scoops with ends arranged in a twist so that they cut into the soil like a drill as the tool was rotated, being a cross between the auger and scoop types (Fig 19T).

Machine cultivation

After ploughing, most soils have to be further broken down before they are ready for sowing. Later, they may need treating to cover the seed or aerating to expose surface particles, they may need compacting, and they may have to be gone over to remove weeds. The appliance to break clods more speedily than by individual hitting was a roller, and the earliest was merely a section of tree trunk (Fig 20A), which may have been bound with iron bands and was sometimes fitted with spikes. A box above the roller became usual to allow extra weight to be added (Fig 20B). There might also be some sort of scraper to clean mud from the roller as it revolved.

Stone rollers were made so as to get ample weight, but they tended to chip and break. Cast-iron rollers were, and still are, the answer to this problem. An improvement with the greater precision in manufacture of these rollers was their division into two parts (Fig 20C). This simplified turning, as one roller went forward and the other backwards on a tight turn, minimizing the amount of skidding round of the longer single roller. Nowadays iron rollers may be towed as gangs or in tandem.

Although two-part rollers are still in use, ring rollers are popular. These may be used in gangs of three, allowing a good width to be covered (Fig 20D), while the assembly can be dismantled or changed to tandem to pass through a gate or along a roadway. Each roller is built up of a large number of rings which are shaped to cut into clods of earth (Fig 20E). The effect is to give a greater pressure than a larger smooth roller and the individual rings, being free to revolve independently, allow of better turning. Apart from making it easier to manoeuvre, this means there is less skidding on the surface, which would tend to tear out small plants, particularly when used on a cornfield.

The ring roller is still often called a Cambridge roller, named after W. C. Cambridge, granted a patent for a ring roller in 1844. His roller had alternate plain and larger, toothed rings which ran slightly eccentrically on the axle—the wobbling uneven motion of the discs proved effective in breaking up clods and had a self-cleaning motion.

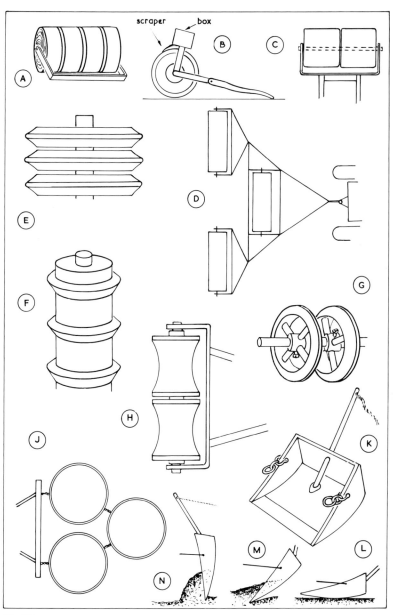

Fig 20

Another ring roller with large spiked rings for breaking up clods had been patented by W. Crosskill in 1841. Crosskill used serrated wheels with sideways projecting teeth. These were particularly successful in pulverizing clods and consolidating soil. Road wheels had to be used to lift the rollers clear for transport. On heavy soil the roller was pulled by one horse in the shafts and one at each side with chains. Both Crosskill and Cambridge rollers were particularly effective on grassland infected with wireworm.

A variation on the ring roller had the fairly thin sharp-edged discs separated by wood spacers. This was used on wet land to promote drainage. The discs made grooves in the soil as deep as the amount they projected from the spacers, which was about 5in (12cm). The device was weighted to make it penetrate, and the discs had to be equipped with scrapers to prevent the gaps clogging with mud. A further development with a family resemblance, although a different purpose, is the disc harrow.

Rollers were made in special forms to suit particular purposes. Ridges on a roller could press seed drills into the earth as an alternative to making them with a cutting tool (Fig 20F). Taking such a tool, with ridges only a few inches apart, over a field after ploughing, could be followed by seed sown broadcast with a good chance that most seed would fall into the drills where it could be covered by using a harrow.

A step up from the seed-drill roller was the use of large, heavy, shaped wheels, adjustable in their positions on an axle. These formed a furrow press (Fig 20G). Pressed furrows were considered better than dug furrows for winter-sown wheat.

For turnips, beet and similar crops, rollers with concave profiles were made (Fig 20H). The roller pressed down the ridge for the seed to be dropped. Later it was incorporated in the seed drill.

Other methods of breaking up clods and levelling uneven ground used weights which slid instead of rolled. Large flat stones may still be found used in this way on the farms of western Ireland. Various cast-iron shapes were used as clod crushers (Fig 20J). Drags, consisting of hurdles, gates and almost anything that could be weighted, were improvised by farmers and pulled across ground to produce a fine tilth.

A device, which probably provided the idea for modern large earth-moving equipment used in civil engineering, originated in the Netherlands as a 'mouldbaert'. It came to England at the end of the eighteenth

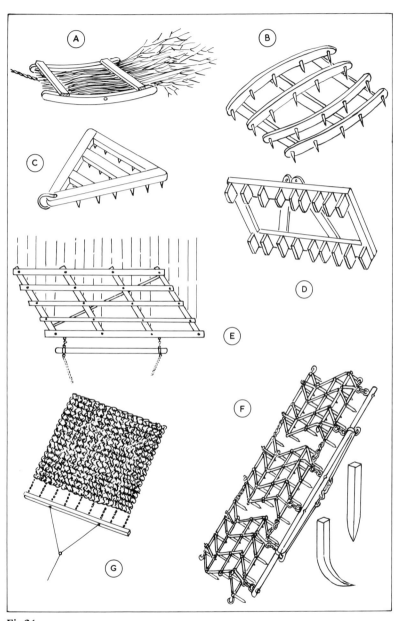

Fig 21

century and became known as a 'levelling box'. Basically it was a scoop-shaped box about 4ft square, sheathed with iron, or entirely iron, on the underside and around a lip (Fig 20K). This was drawn by a pair of horses, with the pull coming from near the point of balance. A single pole at the back was the control used by the driver. Such a box might carry as much as half a ton of soil. The weight provided a crushing action (Fig 20L), but the main use of the levelling box was to take soil from high parts and deposit it in hollows.

To remove a high spot the driver lifted the handle to cause the lip to scoop soil into the box (Fig 20M). Lowering the handle stopped scooping, then at the hollow a quick flip up caused the lip to dig in and the pull of the horses made the device tilt to shoot the soil out (Fig 20N). The lip scraped the pile level as the horses pulled. Pulling the handle back, usually with an attached rope, brought the scoop into position to move the next high spot.

Harrows

Rollers compress the surface as well as break up clods of earth. The tool for loosening and dividing the surface into fine particles is a harrow. The earliest harrow must have been a branch of a tree dragged across the surface, and there are records of elaborations on this at least as far back as Roman days, when spiky branches were drawn together in some sort of frame. Such a type was the Austrian bush harrow (Fig 21A). Assemblies made like hurdles were also used, and early writers mention osiers woven to hold branches to make bush harrows. Harrows of similar form were made and used in England until early in the nineteenth century.

An appreciation of the value of spiky branches must have led to the development of harrows with teeth or spikes built-in as tines. Iron spikes came into use when the material became available, as wooden spikes would not have been very durable. However, they persisted during the Middle Ages and wooden-spiked harrows were used by American pioneers much later, due to the shortage of other materials (Plate 21).

Early harrows were wooden framed and of no great weight, so logs or stones had to be carried on them to keep them in action while being drawn by oxen. Usually the harrow was allowed to follow the oxen as it would, without control by the driver, but some pioneer harrows had handles so that he could control direction and maybe add pressure.

Plate 21   Basic American pioneer harrow made from a natural wood crook (Cade's Cove, Great Smoky Mountains, USA)

Plate 22   Triangular iron harrow (Farmland Museum, Haddenham)

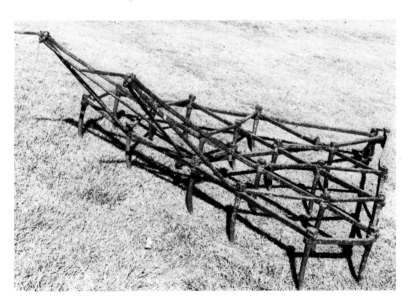

103

Many of the various forms of early harrows were attempts to get the tines following separate tracks and so be more effective. A simple rectangle with spikes evenly placed tends to draw the spikes into line as it is pulled, even when an off-centre pull is arranged. Many harrows were made from bent, cleft wood so the spikes would not be too symmetrical (Fig 21B).

Wooden harrows still used in Mediterranean and Middle East countries are little different from those in use at least 500 years ago. An Italian type has a triangular form, so as to stagger the tines (Fig 21C), while another version has rectangular blocks of wood instead of pointed tines (Fig 22D) to reduce breakages. Both types need weights added to be very effective on heavy soils.

Triangular forms (Plate 22) were used to stagger the tines on some British harrows, but a similar effect with provision for a greater number of tines was obtained by pushing a square out of shape (a rhomboid). As one of these might wander into line as it was pulled, two were fixed together, or the pull was arranged wide (Fig 21E) instead of central. Wood-framed iron-spiked harrows of this type became usual at the beginning of the nineteenth century and there was a transition into completely iron harrows of the same basic design.

The use of tines in a rigid frame does not allow of much improvement and there are present-day iron harrows basically of the nineteenth-century form. A common form is a zigzag (Fig 21F), with tines at each corner of a parallelogram. A patent for this type was granted to W. Armstrong in 1839 and these harrows were first made in Bedford. By the design of the frames no tine follows exactly behind another. The tines are removable and straight for ordinary use, but hook-shaped ones can be fitted for dragging weeds out of the soil. The harrow shown is in three sections flexibly linked. This allows dismantling for transport. The parts follow any slight unevenness of the ground and extra sections can be attached. The broad whippletree keeps the pull straight and this is used behind a tractor in the same way as when it was horse-drawn.

One improvement on the fixed tines was to make them springy. This was achieved by making each tine of steel and giving it a curved form so that as it was dragged forward the pull tried to straighten the curve and make it spring into lumps of earth or ride over obstructions. A patent was granted in America in 1869 for such a spring-tine harrow, but the type did not come to Britain until the end of the nineteenth century. Spring-tine harrows are still used today on some British farms.

Experimenters had ideas about making the tines revolve. Spiked star wheels were mounted on axles, giving something of the same effect as spiked rollers. Morton's revolving brake harrow had a series of star wheels, each with ten spikes or tines. These were set in rows on two axles at an angle to each other and to the direction of pull. The machine was very effective on clay soils infested with couch grass.

Other ways of getting the same effect were tried and a chain link or web harrow is credited to Smith of Deanston, although there were several variations on this by other designers in the mid-nineteenth century. The chain may consist of parts loosely linked to form a web or net. In some forms the harrow parts are smooth and the effect is dependent on weight, while in other forms the links are given spikes or other shapes that break the soil by penetration, with different lengths on opposite sides to allow for a different effect when turned over. The web of links is kept spread by being anchored to a bar by lengths of chain (Fig 21G). These harrows are still widely used.

The spiked star wheels led up to the disc harrow, which is in use on arable farms today. Disc harrows were first made and developed in America. The first patent there was granted in 1847, but a design more like that which became acceptable, and the forerunner of the present-day disc harrow, was dated 1854. With improvements, this has been the form in use for most of the twentieth century.

In a disc harrow there are a large number of sharp steel discs, each of which is slightly concave. The axles are at a slight angle to the direction of pull. As the harrow progresses each disc cuts into the soil and throws it sideways. To avoid the considerable sideways reaction of a single set of discs, harrows are arranged either with opposing gangs in tandem (Fig 22A), or beside each other (Fig 22B).

If the angles of the axles on a disc harrow are adjusted, the degree of penetration and pulverization can be varied, with the lightest working coming when the wheels are running nearly in line with the pull. There have to be ample scrapers (Fig 22C), mounted to clear the discs and prevent a build-up of mud and soil. There also have to be road wheels that can be attached to keep the sharp discs clear when the harrow is moved (Fig 22D).

With a tandem machine the soil is crumbled and pushed to one side by the forward discs, then the second set further crumble it and push it

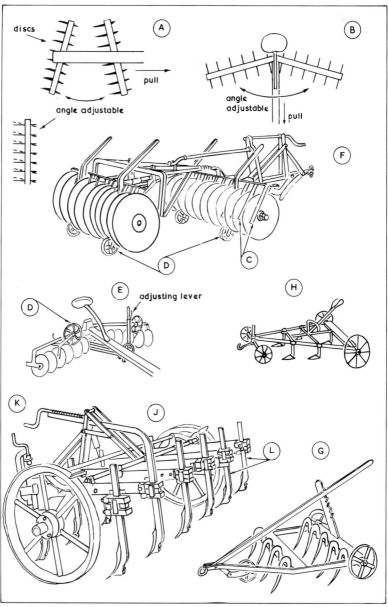

Fig 22

back. On machines with the axles beside each other this does not happen, but with tractor towage a second complete harrow, pushing the soil the other way, can be towed behind the first to get the same effect. With a horse-drawn disc harrow the driver rode on the harrow, where his weight gave some advantage, and used a lever (Fig 22E) to adjust the angles of the axles. With tractor-towing the harrow is dependent on its built-in weight to achieve penetration. A handle to control the angles of the axles (Fig 22F) is brought within reach of the driver.

Cultivators

It was not until the end of the eighteenth century that cultivators began to take over from heavy brake harrows. Earlier devices were known as grubbers, from their ability to grub up roots of weeds. Cultivators differ from harrows in being mounted on wheels so that the tines work at a controlled depth. The tines are usually shaped cutters, rather than mere scratching points.

Much of the early development of cultivators took place in Scotland, where opening up the soil by means other than ploughing may have had advantages. John Finlayson produced a nine-tine grubber on his farm in Ayrshire in 1820. This was entirely of iron and ran on three wheels. The depth of working could be controlled by a lever adjustment of the height of the leading single wheel. The tines curved up before going down and this caused weeds and other things brought up to sweep over the top and clear that way (Fig 22G). An improved version was produced in 1841 at Luey by the Earl of Ducie's farm manager. This was more of a machine, with a high frame and a worm mechanism to adjust the height of the tines.

Another name that helped to set the pattern of subsequent cultivators was Arthur Biddle, who was awarded a Royal Agricultural Society medal in 1840 for his scarifier (another early name for cultivator). This carried two rows of curved tines on a framework between two large wheels, while a pair of small close ones at the front took the pull from the whippletrees attached to the horses.

Later cultivators were equipped with spring tines to minimize breakage and jamming. Some machines were also given a lever action to raise the tines for cleaning.

With a horse-drawn cultivator, the driver walked behind and could make adjustments with one or more levers (Fig 22H). With a tractor-drawn cultivator, the implement could be more massive (Fig 22J). Handles within reach of the driver allow alteration of depth of cut (Fig 22K), or lift the tines clear for transport or turning at headlands. The stalks of the tines can also be adjusted for depth (Fig 22L) or changed. Narrow chisel points will go in deeply, to break up hard land and drag out long weed roots; wide shovel points are used more as hoes near the surface, to cut off weeds; chisel points can be used at a high setting for clod breaking after hoeing; there are also broad ducks'-foot cutting shares that can be used right on the surface for cutting off thistles and similar weeds.

As alternatives to ploughing, several inventors turned their thoughts to digging, using machines that were forms of cultivator. In fact, no means of mechanical digging were able to take the place of ploughs, but there were several devices produced that used rotating cutters to break up the ground.

In the mid-nineteenth century, steam engines were produced with digging devices under their rear ends, but none of these had much success. In 1857, Thomas Rickett put a screw-shaped cutter under an engine, driving it with chains in the opposite way to the motion of the engine (Fig 23A). Robert Romaine, a Canadian, used a spiked cylinder (Fig 23B). Frank Procter, of Stevenage, used forks worked with a reciprocating motion under the rear of an engine, and claimed to work ten acres per day. An even more massive, twenty-ton machine, of which thirty were produced in 1877, was the idea of Thomas Churchman, an Essex man. His engine sat crosswise over four rows of ground wheels. The digging gear consisted of cultivator-type tines that followed the wheels to deal with a piece of ground over 20ft (6m) wide at each pass. Turning at the headland must have been a problem, and, as the machine was too wide to go on the public roads, there was the complication of semi-dismantling to get wheels in line.

More practical was the rotary hoe, credited to an Australian engineering apprentice named Arthur C. Howard. In 1912 he notched the circumference of the blades of a disc harrow, which he drove from a tractor. This threw soil and stones in all directions, but he experimented with blades of various shapes, eventually settling on an L-shape, which is still used (Fig 23C). He built a small rotary hoe powered by a motor cycle engine before World War I. With the mechanical progress during the war,

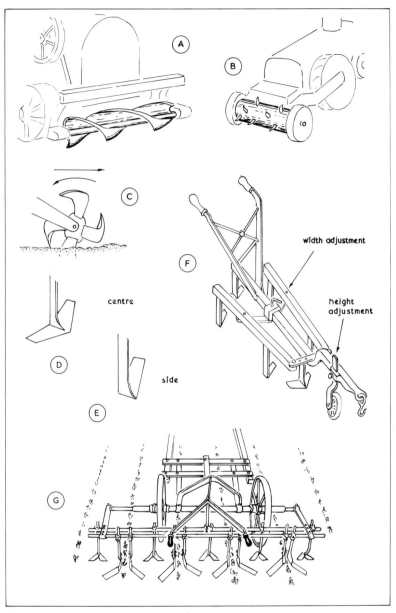

width adjustment

height adjustment

centre

side

Fig 23

he was able to get an American 60hp engine to power a rotary hoe that worked a width of 15ft (4m) and could cultivate over $3\frac{1}{2}$ acres per hour.

In the early 1920s Howard had insufficient capital to market his machines, although he sold some, and he turned to making similar equipment to be driven off tractors produced by other firms, having considerable success in Australia. He produced hand-controlled machines for use in orchards, market gardens and similar places. The name 'Rotavator' was coined and products of the firm of that name are now in use all over the world.

Hoes

Machine hoes and cultivators have much in common and the names tend to overlap. Indeed, in some cases the same machine can be used for both purposes by altering or changing tines. The name 'cultivator' tends to be used for both types of machine in America, but by British definition a cultivator digs deeply, while a hoe operates near the surface. The main purpose of a hoe is to remove and destroy weeds that might otherwise smother a growing crop, although it also does something to cultivate the soil by stirring the surface. Other names that have been applied to cultivators and hoes, and may mean either or both, are 'scarifiers' and 'scufflers'.

Hand hoeing was the only method used for a long time (Fig 19, p 95), but this was slow and must have involved a large number of labourers in a field of any size. Early hoes were given a variety of shaped blades. The type which accompanied Tull's first seed drill had a share something like that of a plough, but the shape that became usual was nearly flat, either on each side of a central stem (Fig 23D), or to one side, for working close to the plants (Fig 23E). There were many variations, but, basically, the tool that evolved for horse-hoeing between two rows had a forward single wheel and two tines at each side of a framework. Control was by handles, similar to a plough (Fig 23F). Gorrie's horse-hoe, introduced in Perthshire in 1840, had a broad transverse blade at the front. This worked near the surface and cut off annual weeds ahead of the deeper blades.

As the worker wanted to hoe as close to plants as possible without damaging them, most hoes included some means of adjusting the width of cut. In early wooden-framed hoes this was a simple peg or

Plate 23 Horse-drawn hoe with width adjustment (Farmland Museum, Haddenham)

bolt arrangement. Later iron models had more sophisticated screw adjustments (Plate 23).

With the width between rows standardized by a multiple seed drill, a multiple hoe was feasible. Some manufacturers produced seed drills which could be converted to hoes—the use of one machine for both purposes ensuring matching spacing. With the fixed position of tines on early multiple hoes, much depended on the steady track of the horse and the ability of the driver to control the whole machine with hoe tines in more than one lane.

This had only limited success and the problem of the horse deviating was later taken care of in the steerage hoe, which allowed the driver to move the tine assembly from side to side to correct tracking. The horse-drawn steerage hoe that developed had the wheels adjustable on their axle, and hoe stalks which could be adjusted on two transverse rods. These adjustments gave the initial setting. The whole assembly of hoes could be moved a limited amount from side to side by the driver, by means of a pair of handles (Fig 23G). With a reliable horse, one man might manage the machine, otherwise there had to be a second man leading the horse.

One of the most efficient, early, horse-steerage hoes that set the pattern for many more was the Garrat lever horse-hoe, shown at the Great Exhibition of 1851. This had fourteen hoes, arranged two per drill space, each on a separate lever so that it could lift over obstructions. The hoes were adjustable and the whole set could be moved by the steerage arrangement. The travelling wheels could also be adjusted, and a lever could lift the hoes clear of the ground.

When tractors became available, steerage hoes were attached to them. At first they still required a man to follow and regulate the hoes to correct any deviation. This limited speed to walking pace, although some hoes had seats, but hoes cutting into plants might have been the penalty for attempting the job too quickly.

CHAPTER 7

Harvesting Cereals

Ever since man has known how to work iron and steel he has used a handled, curved knife when reaping wheat, barley and other corn crops. The design originated independently in many parts of the world and the variations, on what seems basically a simple form, are numerous. Pictures of farmers in Egypt and in Roman days show these reaping knives, which are also mentioned in the Bible. There is a tendency to call them all sickles, but this is not always strictly correct by modern definition.

Most early sickles were comparatively light and small. The blade has always been crescent-shaped, the outline being, approximately, part of an ellipse. The angle the handle makes with the crescent is important for the balance and action of the tool. This was sometimes on the same plane as the blade, or cranked to give clearance near the ground. Early sickles were 12in (30cm) to 18in (45cm) across the blade. This light slender blade necessitated the worker stooping, and only a few stems of corn could be cut with each stroke. These were held in a bunch by the other hand. Early in the nineteenth century better grades of steel became available and there were improvements in the manufacture of these and other tools, mainly due to production being taken from the small smithy to the British manufacturing areas of the Midlands and south Yorkshire, where many British tools are still produced.

Early sickles had smooth blades (Fig 24A), although serrations had been tried. A serrated cutting edge is less likely to slip, particularly on damp straw. Hand reaping was a slow process and improvements in steel and manufacturing techniques brought attempts to make tools that would work quicker. Sickles became larger, their serrations finer, and the steels used retained their edge longer. Fine serrations (approx 300 per inch, 120 per cm), pointing towards the handle, helped the reaper to

113

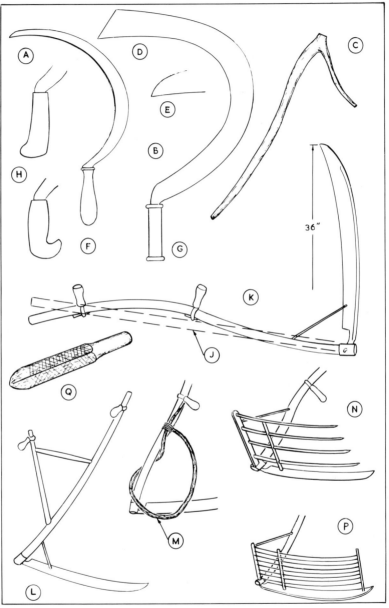

Fig 24

make accurate, quicker cuts. While 'sickle' was the general name, a larger implement became known as a 'reaping hook'.

Before the middle of the nineteenth century, the sickle was giving way to an altogether larger and heavier tool, with the balance concentrated at the end, away from the handle. A weight of 4lb (1.8kg) was possible, being rather more than twice the weight of a sickle. This was generally described as a 'bagging hook' (Fig 24B), although its use had a range of local names, such as 'fagging', 'swapping', 'hewing' and 'cuffing'. Bagging was about four times as fast as using a sickle, but wasteful both of straw and corn. A hooked stick was used to hold a sheaf of corn while the bagging hook slashed at it. This was locally cut for the job and called a 'crock' or 'thank' (Fig 24C).

Reaping with a sickle was considered women's work in the north of Britain. In the south, this and the heavier bagging hook were used by men, but the bagging hook never became popular in the north. Bagging hooks were made in many sizes and angles of curve and manufacturers had to cater for preferences for square (Fig 24D) or round points (Fig 24E), as well as handles ranging from straight and shaped (Fig 24F), or parallel (Fig 24G), to those with ends hooked (caulked) either way (Fig 24H).

Alongside the sickle, at least from medieval times, there have been many forms of scythe. This has the advantage of being used from a standing position and so should be easier on the worker's back than stooping to reap with a sickle or bagging hook. In some places barley and oats were mowed with a scythe for many centuries. Towards the end of the eighteenth century, scythes were used on wheat in southern England, but elsewhere there continued a preferance for the singlehanded sickle or hook.

The curved shaft, still seen in modern scythes, is particularly British. The shaft is described in catalogues by the old word 'sneathe', also spelled snedd, snaith, snath (America). It is important to position the two handles (doles) correctly in relation to the blade. European workers place these on a straight shaft (Fig 24J), while the shaft in southern England has always had more of a dog-leg or S-shaped curve (Fig 24K) than those used further north. Traditionally the doles were on iron rings loosely circling the shaft, to which they were fixed by wedging. Modern tools may have this, or else a screw-locking arrangement. Another way of getting the two doles in the correct position is to have a double Y-shaped handle. This form was usually called an 'Aberdeen' scythe (Fig 24L). Some examples of old scythes appear to have been used with only one dole—the other hand being on the shaft.

Blades were of various lengths (to over 3ft, 1m) and could be changed. Continental scythe blades were much smaller and lighter than British ones and could be used by a woman, but the British scythe has always been a man's tool. Scythe blades mostly had hook ends to fit into an iron arrangement at the foot of a shaft, where security was provided by wedging. Most scythes produced today do not show any improvement on this method. Larger blades were given a strut to a point higher on the shaft, to add some rigidity.

Although a scythe was quicker than a sickle or bagging hook, both hands were used with it and the grain fell haphazardly after it was cut and had to be gathered into sheaves by someone else. An attempt to gather the corn was made by fixing a bow or bale to the scythe. This was a twisted hazel rod (Fig 24M). A step further was a cradle, consisting of four or more wooden prongs, or fingers, braced to the shaft and parallel with the blade (Fig 24N), to catch the grain as it was cut and allow the man to deposit it neatly at the end of each swing. An alternative American cradle, made of light iron rods, fairly closely spaced, must have been heavy to swing (Fig 24P). Good cradlers in America (where scythe cradles were more popular), being paid by the piece, were expected to mow $1\frac{1}{2}$ to 3 acres in a day, depending on the thickness of the stand.

All of these tools needed constant sharpening. A piece of stone might be used and, in recent times, manufactured flat, or oval-sectioned, scythe stones have been available, often carried in a leather sheath on the user's belt and lubricated with spittle. An older alternative was a piece of wood, usually square-sectioned with one end shaped as a handle. This was a 'riff', a 'strickle' or a 'ripe stick' (Wales). It was smeared with grease (mutton or other animal fat) and covered with sand. Grooves across the wood helped to hold the grease and sand mixture (Fig 24Q), and the riff was used like a file on the blade edge. Obviously it could only be used in one direction, from the back towards the edge, or the blade would cut into the wood. Coarse sand or stone would leave an edge with small serrations that were a help in dealing with brambles, while fine sand worked an edge more suitable for hay or corn.

Edge tools on the farm were, and often still are, sharpened by a sandstone wheel mounted on an axle with a crank handle at the end. This was

used with water which, with the necessarily slow speed, had the advantage of not risking drawing the temper of the tool, as can happen with
modern, smaller, high-speed grinding wheels, which overheat the steel
when used by inexperienced workers. Various sizes of sandstone wheels
were used, according to the stone available, but a new stone might be
about 24in (60cm) diameter and 5in (12cm) thick. It wore away in use,
but slight eccentricity did not matter. Some later wheels had a trough
below, but if water was left in this and the stone unused for some time,
the part of the stone immersed became soft and wore quickly. So that one
person could turn the stone and hold the tool, some grindstones were
fitted with treadles.

The sandstone left a fairly coarse edge, with small serrations matching
the size of the grit in the stone. For some purposes this may have been an
advantage, but where a sharper edge was required, grinding was followed by rubbing with a stone and sometimes stropping on leather
treated with a paste of fine grit and grease.

Mechanized cutting

Cutting and gathering a crop by hand was tedious and laborious as
well as time-consuming. Even when labour was cheap and plentiful there
was always the weather to consider, and speed in reaping while the
weather was dry was always a prime consideration.

The earliest record of an attempt at mechanized harvesting is given in
the writings of the Romans, Pliny and Palladius. During the period when
the Romans overran most of Europe, in the first century AD, they
devised a 'vallus', which was pushed into the crop by an ox or mule.
Exact details are not known, but from surviving drawings it appeared to
be a two-wheeled cart with cutters (Fig 25A). The heads of grain were
torn rather than cut and fell into the cart, possibly being helped by men
with rakes or poles, while the straw was wasted underneath. There is no
evidence, however, that these aids to reaping were ever used in Britain.

It was not until well into the eighteenth century that thought seems to
have been given again to reaping machines. Experimental machines
were mostly unsuccessful. Any that worked could only do their job with
the ground level, and the straw in good condition and standing erect.
The first patent for a reaping machine was issued to a Londoner, John
Bryce, in 1799.

117

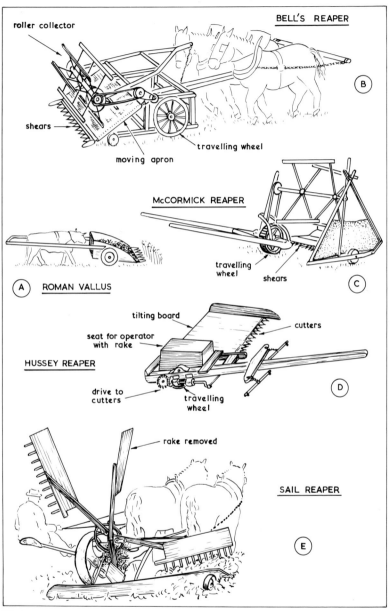

Fig 25

Early inventors tried to make their machines work like scythes. Donald Cummings received a patent for a revolving knife reaper in 1811. The same year, James Smith of Deanston produced a reaper with a continuous revolving knife. A large-diameter steel-cutter was arranged to turn horizontally a short distance above the ground, taking its power from the travelling wheels. The whole thing was pushed by a horse. Above the cutter, but not of such a great diameter, was a canvas drum. The cut corn fell against this and was thrown to the side. Smith claimed that his machine would cut one acre per hour, and during that time the knife would have to be sharpened four times. To assist in laying the corn when cutting in opposite directions, the cutter could be adjusted to turn either way. However, the machine was unwieldly, especially when changing direction. If the land was uneven and the knife cut into a ridge of soil, it was blunted.

The first reaper using reciprocating shears was patented in 1807 by Samson, of Woburn. This cut the corn with a clipping action in the way used in many other machines that followed, but Samson's own machine was not successful.

Some people had a superstitious fear that harvesting by mechanical means was working against nature, and some agricultural workers opposed mechanization as likely to cause unemployment. Many experimenters had to work secretly. Credit for the first successful nineteenth-century reaper goes to the Rev Patrick Bell of Carmylie, Forfarshire, in 1826. He is said to have made his first trials in a barn, using a crop planted stalk by stalk.

Bell's reaper was pushed by horses on each side of a pole with a handle at the end, by which the driver steered, and the machinery operated from the two large travelling wheels. Construction was an open wooden framework and cutting was done by a series of thirteen double-edged shears, with the moving shears rod-operated. A six-bladed revolving-roller collector forced the cut corn onto a moving apron, which turned it into a swath alongside (Fig 25B) and could be adjusted to throw either way. The cut was 5ft (1.30m) wide and an average performance of one acre per hour was claimed. Bell was awarded a premium by the Agricultural Society of Scotland in 1827 for the reaper, but refused to patent it as he wanted it to be of general benefit, and so others made copies, often unsuccessful.

Although Bell's reaper was the most workable up to that date, only a few were built and actually used in the field on crops. It is believed that four machines were exported to America, and the further development of reaping machines then moved to that country, but there was no significant progress until the middle of the century, when the names of Hussey and McCormick came into the story.

Cyrus McCormick of Virginia, USA, is credited with inventing and building his reaper on his farm at Steele's Tavern in July 1831. It is thought that he must have been influenced by the earlier work of his father, Robert, and there are points that could be regarded as being very similar to the Bell machines. However, Cyrus had the enthusiasm and commercial ability to make a success of spreading the use of his machine. He obtained patents in America in 1834, and started selling in 1840. As a result of experience, improvements were made and patented in 1845. The flood of men from the land to take part in the California gold rush of 1849 gave a boost to sales. The McCormick reaper came to Britain for the Great Exhibition. By then snags had been smoothed out and this was a practical harvesting tool (Fig 25C). McCormick, by then working in Chicago, received a gold medal from the American Institute for this machine. His business methods were such that, by 1860, his annual sale of the reaper was 40,000.

McCormick laid down seven basic principles for his machine:

1 The straight knife with serrated edge and reciprocal or vibrating motion, which cuts the grain.

2 Fingers or guards extending from the platform to prevent the grain from slipping sideways while being cut.

3 The revolving reel which holds the grain against the knife and lays the cut stalks on the platform.

4 A platform behind the knife for receiving the cut grain and holding it until raked off.

5 The master wheel, which carries most of the weight of the machine and, through ground traction, furnishes power to operate the reel and the knife.

6 Forward draught from the right or stubble side by means of shafts attached in front of the master wheel.

7 A divider on the left side to separate the grain to be cut from that to be left standing.

Also at the Great Exhibition was an American Hussey reaper,

patented in 1833. This too was pulled by horses at the side and the multiple shears, consisting of triangular blades reciprocating between two lines of projecting guards, extended to the left of the operator, who was able to travel on the machine, sitting on a case over the main travelling wheel behind the horses, which were driven by another man. There was no reel, as in the other machines, and no proper scissor action. The operator used his rake to control the grain over the cutters and passed it off the back of the machine onto the ground by tilting the board with his foot (Fig 25D).

Following the Great Exhibition, the McCormick machine was made in Britain under licence by Samuelson's of Banbury, while Dray of London improved and built Hussey-type machines. It was claimed that 1,500 Hussey reapers were sold in Britain in the two years after the Exhibition. By 1869 there were eighty-four types of reaping machine taking part in the Royal Agricultural Society Manchester trials. It was esti-

Plate 24   Sail reaper pulled by a horse in the shafts and another beside it (Jackson Collection)

mated that a quarter of Britain's corn crop was being mechanically harvested by 1871.

A machine that needed a man to rake off the cut grain was replaced in the later 1850s with one that delivered the grain automatically to one side. This was done by four revolving rakes or sails (Plate 24), which took their power from the travelling, or bull, wheel and scraped the grain off at fairly frequent intervals, leaving it in regular bunches on the ground (Fig 25E).

Of course, it still had to be gathered, bound and stooked by hand. In some machines the rakes could be removed from two opposite sails if larger bunches were required. There was no provision for a driver to ride on some machines, so he rode one of the horses, adding further to their load. Later machines provided a seat for the driver, usually in a position where his weight helped the traction of the travelling wheel. Other means of discharging the grain were tried, but none equalled the revolving sails, which remained the usual method into the twentieth century. As an alternative to removing rakes to get larger piles on the platform before pushing off, the gearing mechanism could be adjusted so that certain sails lifted and only alternate or every third sails swept right round. Pulling the machine was only part of the load on the horses, as they also had to provide the power for the machinery from the ground wheel(s), so frequent changes of horses were necessary.

Hand gathering tools

When wheat, barley and oats were harvested with a reaping hook or scythe, and even when machines had advanced to cutting and depositing the grain on the ground, there were follow-up processes, involving many workers, with special tools. Large rakes were used to gather the crop. These were usually made as light as possible, from softwood, instead of the usual ash or other hardwoods, so the largest tool could be used with the minimum effort, often by women. Further lightening was obtained by bevelling and rounding the wood. In such a 'drag', 'hobby' or 'ell rake' (Fig 26A) there was a width of about 5ft (1.50m), and iron tines about 3in (8cm) apart were generally preferred to wooden ones.

Rakes varied between counties. 'Stail' was a common name for handle. Some had the spread part of the handle shorter (Fig 26B). Rounded wooden tines were more often used for hay, while a binding

122

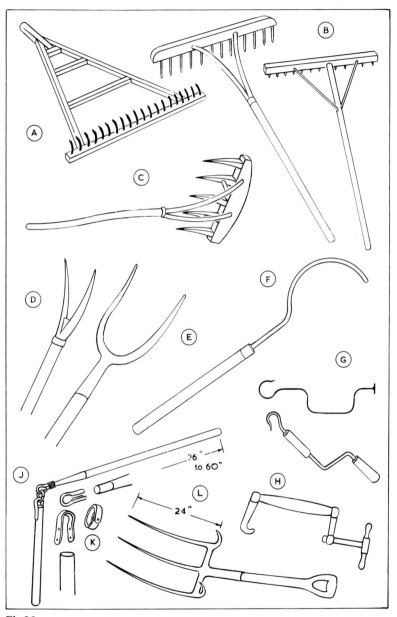

Fig 26

rake had long curved tines (Fig 26C). This was dragged until full when it could be turned over to hold the sheaf while it was bound. Horse-drawn rakes were also used. These could be lifted by the driver so that the corn was left at intervals, in bunches of a suitable size for sheaves. More sophisticated versions (Fig 30, p 135) may still be seen gathering hay behind a tractor.

The pitchfork had uses in the harvest field as well as in yard and barn. Earlier versions were made of wood. A split end of a rod was held open by a wedge and prevented from splitting further by a leather band (Fig 26D). The steel tool, more general in Britain and still available (Fig 26E), was used mainly for tossing hay and straw onto a waggon or onto a rick being built. This job is now done mechanically by an elevator, but there are still uses for a pitchfork on most farms. Natural, wood hooks were also used to gather together the cut grain. They would be cut as needed, or a handled iron 'gavel' might be used (Fig 26F).

In the days when sheaves had to be gathered, bound and stooked in the field, straw was used for binding. Lengths of straw were twisted into rough ropes. To facilitate this twisting there was a tool, operated rather like a carpenter's brace, but simplest was a cranked iron rod, which might be improved with wooden handles (Fig 26G). There were all-wood ones (Fig 26H) and a variety of metal shapes. Some of the many names given to this tool are: 'hay bond twister', 'wimble', 'whimble', 'scud winder', 'throw hook'.

Self-binding reaping machines

Reaping machines gained considerably in speed over hand reaping, but in most cases the number of men needed to get the crop in was still quite high. More than one man was needed on the machine and many more were needed to follow on and deal with the cut corn, binding and stooking it. The earliest machine that would both reap and bind the sheaves was supposed to have been made by a Northamptonshire school-master in 1822, but this was not produced commercially. Experiments began in 1850 in America, and in 1878 the McCormick company started marketing a self-binding reaping machine. There had already been rea-pers on which a man travelled and bound sheaves by hand; a British self-binder was made by Hornsby in the 1880s.

The McCormick machine was a reaper working in the same way as the

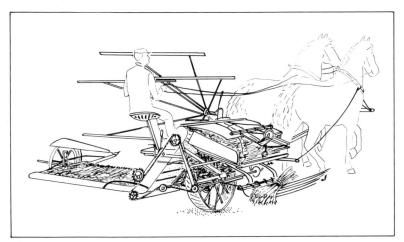

Fig 27

earlier machines, but it deposited the straw automatically tied with
string. This meant that another step in the labour chain was removed,
although men still had to stook the corn, and the field had to be raked,
either by hand or horse.

The McCormick idea was followed by many other manufacturers and
string-tying self-binders have continued in large numbers almost to the
present day (Fig 27), although combine harvesters have replaced most of
them, particularly for dealing with large fields. Self-binding reapers appeared at about the time when steam and internal-combustion tractors
were coming into farm use, so their use and development were complementary. The usual horse-drawn self-binder needed three horses to pull
it (Plate 25). With mechanical power, wider cuts were possible, making
fewer passes over a given area to complete the job and a 7ft width of cut
on a self-binder was then common. Early binders were towed and it was
sometimes possible to tow two self-binders. Self-powered binders came
much later. Obviously, the use of power which could be disconnected
and used for other purposes was a better proposition for the farmer.

Towards the combine harvester

The reaping machine, progressing to the self-binder, was the limit in

125

Plate 25   A Massey-Harris binder, about 1900 (Rutland Museum)

harvesting development in the field for a long time. Corn still had to be threshed elsewhere. In Britain this was usually a winter job with flails. The corn was stored in a barn via a central entrance, through which a waggon could be pulled and the straw unloaded to either side. In the winter this central area became the threshing floor, on which a group of men used their flails to beat out the grain.

A flail consists of one pole hinged to another one by a universal joint (Fig 26J). The handle, like those of most other tools, was ash, but the favoured wood for the beater (swingle) was holly or blackthorn—both hard, heavy, close-grained woods. This was about half the length of the handle, hence the colloquial name of 'stick and a half'. The joint was made from a piece of yew or ash, steamed to shape and lashed on to take a leather link (Fig 26K).

126

In other countries the grain was, and often still is, spread on the floor and threshed by animals drawing heavy sledges over it. Oxen tethered to a central post, with women riding on the sledges to provide extra weight, were seen by the author in Turkey recently (Plate 26).

An alternative to beating ears of corn with a flail was to reverse the process and beat the ears against a frame of crossbars so that the grain would fall through. This was done when unbruised straw was wanted for thatching. The semi-drum arrangement of bars may have provided ideas for the machine threshing drum.

Waste broken straw was called 'covings' or 'colder'. This was gathered with a long coving fork (Fig 26L) with two or three prongs, and hooks above the outer prongs so that a bunch of straw could be retained on the fork and taken elsewhere.

Attempts at threshing by machine came before the mechanization of most other farm processes. Sir John Christopher van Berg received a patent for a threshing machine in 1636. This used a number of revolving flails. Several others followed and, in 1758, Michael Stirling used the

Plate 26   Threshing grain in Turkey with oxen pulling heavy sledges on which the women are riding

idea of a drum rotating within a cylinder. Credit for the first threshing machine with a useful work output was given to Andrew Meikle, who built it in 1786 on the farm of a Mr Stin of Kilbagie. Dissatisfied with earlier machines, he designed his own, in which the ears of grain were beaten by being passed between a rotating drum with lengthwise bars and an outer casing—an idea unsuccessfully anticipated by Stirling, but which was used in many later threshing machines.

Demands for more corn, with which the flail's output could not keep pace, obviously gave an incentive to the designers of threshing machines. The population was concentrated mainly in the southern half of Britain and machines first found a place in the north, where manpower was less. In the south, workers, who saw the machines as a possible cause of winter unemployment, indulged in machine wrecking and it was not until nearer 1840, when the workers were assured of winter work, that threshing machines came into use in the south.

Early machines in Scotland and northern England were water or horse powered, but the smaller machines in southern England were hand operated. A Ransome's hand threshing machine of about 1840 was operated by four men and could deal with upwards of one ton per day. With the coming of steam, static steam engines were used to drive thresh-

Fig 28

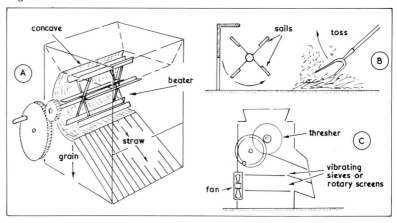

ing machines from about 1840 onwards. Horses were used at least until the mid 1860s.

Threshing machines, of the pattern that became accepted, remove the grain from the ears by a revolving drum. In the early machines, sheaves had their bands cut and were fed by hand into the top of the drum box. Bars or beaters on the revolving drum knocked the grain out of the ears against a concave grid. The grain fell through the grid and fed out through a chute. The straw continued on to be pushed out at another point (Fig 28A).

In its basic form this was all that the early threshing machine did. The grain was mixed with chaff, weeds and other unwanted things. Removing these impurities was first done by tossing the grain in the air (winnowing), often from a broad shallow basket. This might be done outdoors in settled weather, but a breeze was needed. The grain fell, while the lighter chaff blew away. In the unsettled British climate, the two doors of the normal barn allowed winnowing on the threshing floor while any draught was directed through. One of the earliest attempts at mechanization was to provide the draught artificially with a home-made fan, consisting of canvas sails on a windmill (Fig 28B).

A step forward was known as a 'fantackle and chogger.' The draught was provided in the same way, but the seed was loaded onto a slightly sloping tray, which shook so that the seed moved downwards and eventually fell off. This was the chogger. While the seed was moving the fantackle blew away chaff and light weed seeds.

Winnowing machines had been introduced quite early in the eighteenth century. James Meikle imported one from Holland in 1710. The principle was that of the fantackle and chogger, cased in. As the effort needed to work a hand-operated winnowing machine was not great, and the results were both better and quicker than hand winnowing, this mechanization found early and general acceptance.

An obvious combination of jobs was to include winnowing in the threshing machine, so that the grain falling from the threshing drum could be sieved and subjected to a draught to winnow it before it passed out of the machine (Fig 28C). At the Great Exhibition, there was Garrett's portable horse-drawn threshing machine, which was probably the biggest British threshing machine built before the coming of steam power. The wooden-cased machine could be pulled by horses, but in use it opened, to be fed from the top, while power was provided by four horses walking around a vertical shaft, driving another shaft to the machine. As in later machines, the grain was beaten out between bars on a rotating drum and a fixed concave. Grain and chaff dropped through a built-in winnowing machine, while straw and unthreshed short ears were ejected at another place. An output of sixty bushels per hour was claimed. A similar type of machine, powered by eight horses, was claimed in 1860 to thresh three hundred bushels a day—without quoting the length of the day.

Later machines, sometimes called threshing boxes, became bigger, particularly as steam and oil engines came into use for driving them, to give them a greater capacity and to add refinements, particularly in the final dressing of the grain (Plate 4, p 37). In such a machine, with its apparent complexity of belt and chain drives, the grain passes through riddles of various sizes, to remove weed seeds, and is blown to remove chaff at more than one stage. In more recent machines a final stage passes the grain through a rotary sieve, which divides it into three sizes before ejecting it from several spouts into sacks or other containers (Plate 27). In the later large boxlike machines, the straw, in its shaking process

Plate 27   A 1903 threshing machine restored to working order

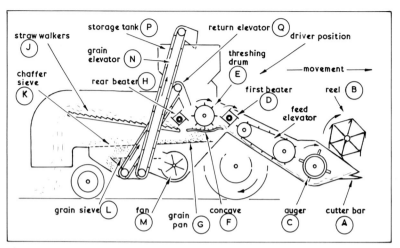

Fig 29

to remove more grain, is also lifted by a reciprocating elevator so that it passes out at a suitable height for loading into a truck or waggon. In some parts of the country the straw that left the threshing machine was called 'coving' and this was gathered up with a hooked coving fork (Fig 26L, p 123).

There are still many threshing machines capable of doing their job, but most stand idle, as the combine harvester has taken over. Steam-driven threshing machines of a century ago, not vastly different from the machines of the recent era, dealt in one week with what might have taken the same number of men, as those tending it, maybe six months or more by hand methods. The vital parts of the threshing machine became part of the combine harvester.

If broken down into its basic operations a combine harvester is really a reaper which takes its thresher with it (Fig 29). The latest machines are self-propelled and have the operator in an air-conditioned cab, with knobs and levers controlling all the operations. The grain is cut, threshed and dressed, and carried on the harvester to be passed out in bulk to another vehicle alongside.

A knife (Fig 29A) on the harvester precedes a reel (Fig 29B), which can be adjusted to suit the crop. An auger (Fig 29C) or elevator leads the crop to the first beater (Fig 29D) and into the threshing drum (Fig 29E), from

131

which the process continues in a similar way to that of a large threshing machine. The drum rotates above a concave (Fig 29F), where bars beat out the grain, which falls through onto a grain pan (Fig 29G).

The rear beater (Fig 29H) strips straw and directs it onto the reciprocating bars of the straw walker (Fig 29J), with arrangements for any remaining grain to fall through.

The grain pan shakes the grain, which may be contaminated with trash, onto a chaffer sieve (Fig 29K), where grain falls through and most trash falls out of the machine. The grain sieve below (Fig 29L) serves as a second stage, while a fan (Fig 29M) blows air through to prevent the lighter trash falling down.

The grain is fed by an auger to the bottom of an elevator (Fig 29N), which takes it to a storage tank (Fig 29P) or a bagging platform. Trash may be returned to the threshing drum via another elevator (Fig 29Q) for a second treatment.

It is debatable whether the first combine harvesters were produced in Australia or America. Both countries had the problem of vast areas of grain to be dealt with. In 1845 the Australians devised a 'stripper' or 'header' type of harvester, which only removed the heads of the grain and left the straw standing. The Americans favoured cutting low and having a threshing drum large enough to take the whole straw.

An experimental machine was built in Michigan, USA, in 1836, and by 1843 it could cut and thresh twenty-five acres per day. Michigan had a damp cool climate and there was considerable spoilage, due to the high moisture content of the grain, when there was no satisfactory way of further drying. The drier western areas did not produce this problem and a machine shipped round Cape Horn to San Francisco, California, cut 600 acres of wheat in 1854. Factory production of combine harvesters started in California in 1880, and by the end of the nineteenth century two-thirds of the California wheat crop was harvested by combine.

These early machines were hauled by mules or horses and all of the power to work them came from the travelling wheels. A contemporary picture of one of these American machines at work shows a team of forty mules pulling it. Steam power replaced the horses or mules. Later, petrol engines were mounted on the harvester to drive the machinery, but the harvester still had to be towed along the ground. Eventually the petrol engine was made big enough to propel the harvester as well. Massey-Harris of Toronto claim to have pioneered the self-propelled combine harvester, developed mainly in the Argentine by Australian-born Tom Carroll.

At the end of the nineteenth century, when combine harvesters were finding a place in American farming, British labour was plentiful and cheap compared with American conditions. With smaller British farms, in any case, the amount of grain was not usually enough to justify changing from reaper and thresher, and the smaller fields were less suitable. In the 1920s the threshold at which a combine harvester was justified was about 300 acres. By the outbreak of World War II there were only about 100 combine harvesters in use in Britain. With changing conditions the threshold has come down to about 100 acres of grain. However, the combine harvester is not the only expense. In the usually damp British climate there also has to be grain-drying equipment to reduce the moisture content to an acceptable figure of about fifteen per cent.

A snag which the modern self-propelled combine has overcome is the need to cut a way around the edge of a field for a towed reaper or combine, which had its cutter bar at the side. A self-propelled harvester can be driven straight in. Clearing around the many small fields of the average British farm would have meant a lot of work, probably with a scythe, before the benefits of the machine could be realised.

The mechanization of the British corn harvest may be reckoned to have started around 1850. By about 1875 half the harvest was got mechanically, until around 1920 when mechanization became just about 100 per cent. Reapers and self-binders took care of most of this progress and the numbers of combine harvesters were quite small until after World War II, when combines more suited to British and European conditions came in, and progress since has been steady, from 10,000 in 1950 to well over 60,000 twenty years later.

133

CHAPTER 8

Harvesting Other Crops

Grass grown for stock feeding is a major crop. A cow grazing in a field is harvesting the crop directly to turn it into meat and milk. However, harvesting grass is normally taken to mean cutting and processing the crop so that it will keep for future use. Hay is usually harvested first and may be an urgent job. Until the coming of silage-making, and other ways of treating grass after cutting, the farmer was very dependent on the weather. The interval between the day when the grass has grown to sufficient bulk and the time it becomes woody and much less valuable is quite short. In traditional grass harvesting, periods of dry weather had to be used to the limit when they came between these times.

For most of history haymaking has been a manual job, with horses only used for transport. Hay was cut with sickle or scythe (Fig 24, p 114), then turned to dry by forks and other hand tools. Hay on the field, called swath (swathe), would only dry on top if left undisturbed. Turning hay by hand took a considerable amount of labour. Rakes of many sorts (Fig 26, p 123) were used to gather the hay, which was loaded onto carts and waggons, then taken to a yard for storage in a large number of thatched ricks. This series of steps in the making of hay has only gradually altered with the coming of mechanization. Many farms still have rick yards, although they are mostly used for other things. The thatching of ricks is almost a lost art and the hay is now baled and stored in barns, either in roof-only 'Dutch' barns, or ventilated, more enclosed buildings.

Mowing

Grass cutters or mowing machines are closely related to machines used for cutting corn, but tend to be simpler. Early attempts at mechan-

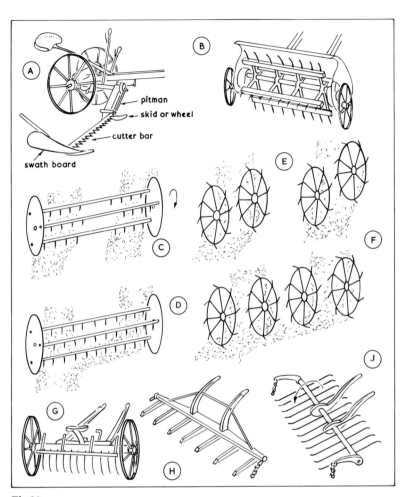

pitman

skid or wheel

cutter bar

swath board

Fig 30

izing grass cutting used swinging and rotating knives (like the rotating-knife reaping-machine by Smith of Deanston in 1811), but the design which proved to be the forerunner of the majority of successful production machines was by an American named William Manning, in 1831. Manning used a reciprocating cutter, working with a multiple scissor action between guards. This had applications for corn harvesting as well as grass cutting, but specialist grass-mowing machines went into production in America early in the nineteenth century and were soon in use in Britain.

In an early mowing machine (Fig 30A), the cutter projects to one side of a two-wheel carriage, pulled by a pair of horses, with a seat for the driver and two or more control levers within his reach. Power comes from the two wheels. Weight is an advantage to make the ground wheels grip and they usually have ribs cast on the rim to reduce slipping. Bevel gears transmit the drive, and in most later machines the drive was taken by pawls and pinions, so if a wheel stopped, or reversed in turning, the pawls clicked over the pinions. This avoided the complications of a differential gear.

A crankshaft and pitman (connecting rod) converts the rotary drive to a reciprocating one. The cutter bar extends four feet (1.25m) or more and is built up. There are guards which carry ledger plates that form half of the scissors. Inside the guards the reciprocating motion moves a knife bar backwards and forwards. The knife bar has sharp triangular knives moving across the gaps between the ledger plates to cut off the grass.

The outer end of the cutter bar has either a wheel or a skid to control the height of cut, and some machines have a wheel at the inner end as well. The outer skid is extended backwards by a swath board to direct the cut grass inwards as it is cut, to fall into a compact swath. A swath stick may be attached to the swath board, for controlling it. On most machines the cutter bar can be swung upwards for transport by pulling a lever and this action also disconnects the drive.

In some modern mowing machines there has been a return to rotating cutting discs as first used by Smith. Such a disc mower is mounted on the tractor, with a cutter bar extending from the side and four discs driven from the tractor rotating at high speed (over 3,000rpm). Modern materials and easily replaced discs avoid the frequent halts for sharpening necessary with Smith's machine. Speed and close cutting are advantages claimed for this type of mower.

Tedding and gathering

A tractor mower has the cutter bar extending from the side or rear of the tractor and drive comes from its power take-off, but the cutting action on the ground is the same as in horse-drawn machines.

The grass, whether cut by hand or machine, has to be shaken up and

136

opened to admit light and air, particularly any breeze. Known as 'tedding', this used to be done by large numbers of women with rakes and forks, tossing and spreading the grass to separate and fluff it up. The spread grass was called a 'windrow'. A simple, early, mechanical hay tedder consisted of a sort of wooden cage on a geared axle between the road wheels, with iron spikes or tines to toss the hay into the air (Fig 30B). Later variations had the cage divided and sprung so as to be better able to cope with uneven ground.

Another horse-drawn version, the American hay kicker, emulated the women with the forks. As the machine moved forward, a reciprocating action was given to a series of forks which dug into the swath and threw it.

A tedder is necessary when the hay has become compacted by wet weather. A less violent action is given by the swath turner, which follows the tedder, or takes its place and turns the grass over to dry the underside. The swath turner as used today is an invention dating only from the early twentieth century. Swath turners may use rake bars or finger wheels. Both types can be set to turn the swath and retain it in rows, or as side-delivery rakes to bring two swaths together.

A swath turner with rake bars has tines on bars between two wheels, driven from the ground wheels, and adjustable so as to turn at varying angles to the direction of travel. The tines follow an elliptical path and remain vertical in all positions (Fig 30C). The tines from the centre parts of the bars can be removed, then those groups at the ends turn the grass over and leave it in approximately the same place for drying. When the whole set of tines is used, the grass progresses across from one row to the other as it is turned (Fig 30D). By halving the number of rows in this way, the labour of picking up is reduced, so this is done just before collecting.

The other, and now more common, swath turner has four or more finger wheels. These have projecting tines and are turned by the tines touching the ground. The wheels are set at an angle to the direction the machine follows on the ground wheels, in a similar way to the wheels of the rake-bar swath-turner. Each finger wheel is independently mounted, but they can be arranged in pairs for swath turning (Fig 30E), or brought into line with each other for side-delivery raking (Fig 30F), just before picking up.

If the stems and leaves of the grass are bruised, sun and air can dry it more quickly. Roller crushers are used and may be arranged to be towed at the same time as the mower, or incorporated in the same machine. While one swath is being mowed, the previous one may be bruised. In the roller crusher, two mangle-like rollers rotate at high speed as the grass is fed between them. Variations have grooved rollers, or interlocking ribbed rollers may crimp the crop, which is supposed to leave the grass more suited to allowing air to pass through.

Another step in the grass-drying process is raking it into swaths, particularly if, in tedding, it has spread unevenly. A horse rake (dump rake) was used across the direction of mowing to gather grass into reasonably straight lines. The usual horse rake had a very large number of curved tines supported on a carriage between the wheels and a seat above for the driver (Fig 30G). In use the operator drove the rake until a load was gathered, then pressed a pedal to lift the tines and deposit the load. By doing this at regular intervals, during each pass across the field, rows were formed.

Another implement which achieved the same result and was used before the iron tool described, was made of wood, in many individual patterns. Some were really large, raked scoops, fenced to hold enough hay. Others were more like sliding broad forks. A type popular for some time, and overlapping the use of the horse rake, was a hay collector or sweep (Fig 30H). This was drawn by one or more horses, with the pull coming via chains from the ends of the main crossbar. When the sweep was full, the driver pushed the handles up so that the contrivance tipped forward to deposit its load, then it turned a complete circle to come up empty, ready to start again.

A similar principle was used in what was called an American rake or sweep, although it is claimed that the idea had already been used in Britain where it was imported in the middle of the nineteenth century. In this form, the prongs extended both ways (Fig 30J). The handles locked against the upward prongs, and when one side had been filled, a sideways movement of the handles released the main body so it rotated half a circle to deposit the load and put the other row of prongs down to start gathering hay. They restarted gathering quicker than the single-sided type and so left less loosely scattered grass.

While rakes and sweeps might also be used finally to gather hay as well as get it into swaths, larger flat sweeps were used to gather hay to be loaded and transported away. Early sweeps were simply boards to pull

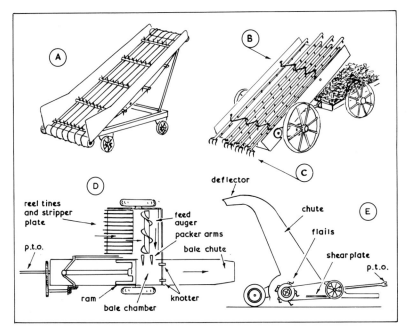

Fig 31

hand-gathered hay across the field. Sledges were also used and these have continued in present-day usage for collecting and transporting baled hay. One method of gathering was to make a fairly large hay cock, by hand-gathering, on a sledge and tow it by horse with a chain around the whole hay cock, to where the rick was being made.

Elevators and loaders were produced before the coming of steam or the internal-combustion engine. Horse power for a static elevator was provided by a horse walking in a circle to drive a sort of capstan that rotated a shaft through bevel gearing (Fig 1, p 18). The elevator, mounted on its own wheels, could be adjusted to accept the load from a waggon or sweep, and have its angle altered as the stack height increased. In the simplest form, the hay or straw was forked onto the bottom of the elevator. The usual method of travel was provided by an endless belt system, like many modern industrial conveyor and passenger escalators, but with cables over rollers and a series of crosswise slats carrying upward-pointing tines. At the top of the chute these turned over to deposit

139

the load and return empty, below the machine (Fig 31A).

The alternative was a series of bars carrying tines and given a reciprocating motion by cranks (Fig 31B). These moved the straw up in a series of stages—as one group of tines disengaged, another took over.

While the stationary elevator had its uses at the rick, thought was also given to combining the elevator with a pick-up arrangement, so that it could be towed across a field and used to load a waggon. One of these was the Keystone type, first produced in America, which showed the influence of the pitchfork. As the loader was drawn along by the waggon to which it was hooked, a series of forked teeth rotated with the ground wheels, which also drove the endless belt type of elevator. Straw was forked off the ground and deposited on the elevator.

Many other loaders followed during the last quarter of the nineteenth century. Some used the general principle pioneered in the Keystone loader, but one produced by the Deere company used the reciprocating-bars idea for the elevator and employed the same idea on the ground to rake the hay onto the elevator. The rakes were attached to the bars and shaped, so that they gathered the hay as the machine moved forward and lifted it on to the elevator bars. While one set of rakes was doing this, the other set came down into position for the next batch of hay (Fig 31C). The bars worked in an open web of fixed slats and the hay travelled between the bars and slats. Motion came from cranks, chain-driven from the ground wheels.

Hay loaders continued in use in the field until pick-up balers took their place in the early 1950s. Elevators are still used, either as separate units or as parts of combined machines. Besides loose straw and hay, elevators lift bales and sacks. The straw walker in a combine harvester uses the reciprocating-bar idea (Fig 29, p 131).

Modern grass harvesting for animal feeding has followed two lines: hay may be baled in the field, or gathered green and made into silage. For pick-up baling the hay has to be cut and dried in swaths and windrows.

Simple hand balers were devised in the nineteenth century. Cut hay was pressed between heavy frames by a long hand lever, then tied. In America, horse-driven balers were devised. In the many presses produced, the power was used to get maximum compression and wire or twine was tied around by hand. All of this sort of baling was done after the hay had been gathered. The pick-up baler, which gathers the hay from the windrow in the field, and deals with it while being towed by a tractor or combine harvester, is a fairly heavy and comparatively new machine. The first pick-up baler was claimed to be the invention of a farm boy who assembled one from secondhand and handmade parts at Kinzers, Pennsylvania. The idea was adapted and went into production at nearby New Holland in 1940.

In one typical modern pick-up baler (Fig 31D), a row of broad wheels or reels carry tines which gather up hay from the swath and lift it to a feed auger. This moves it sideways into the bale chamber, where a crank works packers and a pistonlike ram moves backwards and forwards to compress the hay. A knife moving with the ram cuts off the end of the bale, and an ingenious device knots twine around it. The complete bale passes along the bale chamber and is discharged via a chute at the end. The drier the hay, the tighter it can be packed.

If a sledge is towed behind the baler, groups of bales can be deposited in the field, making collection easier than if they were scattered. Some balers have a throwing device to toss bales into a following trailer, as they are ejected.

Grass cut green for silage is today dealt with by a forage-harvester, a comparatively recent machine which came into use commercially in

Plate 28　A Bamford forage harvester cuts, pulverizes and blows the grass into a trailer

1958. A forage-harvester cuts the grass, chops and pulverises it, and passes it, via a chute, into a trailer, to be taken away and used green or made into silage.

The forage-harvester is usually driven by the power take-off of the tractor, and towed on a pair of wheels (Fig 31E), although there are self-propelled machines. Flails whirl around at high speed towards a shear plate. The grass is cut, broken and pulverised into tiny pieces, which are blown up a chute into the trailer (Plate 28). By adapting and changing attachments, it is possible to alter a forage-harvester to mow or chop hay, clear scrubland and cut other crops.

Potato harvesting

Potatoes were grown for food in South America long before they came to Europe. Brought to Spain in the sixteenth century, they soon found their way into Ireland where, by the beginning of the nineteenth century, they became the staple food. The story that Raleigh brought the potato to England from Virginia cannot be true, as potatoes were not grown in Virginia at that time. There are records of potatoes being grown as a commercial crop in England near the end of the eighteenth century.

Harvesting potatoes by hand was obviously a slow process, so a plough with a special share which could turn over the earth containing the potatoes—doing much the same job as a hand fork—was soon in use. This only left picking up the potatoes to be done by hand. A potato spinner also lifts both earth and crop, leaving the potatoes to be picked up by hand. In the earlier and more basic form, the horse or tractor-drawn potato spinner was supported by two large travelling wheels, which provided the power to work the spinner, mounted on a lengthwise shaft (Fig 32A). Ahead of the spinner, a share was set to cut through the earth below the potatoes (Fig 32B). This loosened the soil and potatoes so that the prongs on the ends of the revolving tines could throw them both to one side.

The success of a potato spinner depended to a certain extent on the type of soil, and later developments were concerned with adjustments and variations to suit differences in soil. Modern spinners are powered from the tractor and any ground wheel is there only for support. The simple spinning tines arrangement may spread both potatoes and earth over a considerable distance, but in heavy soil this may be the only way to

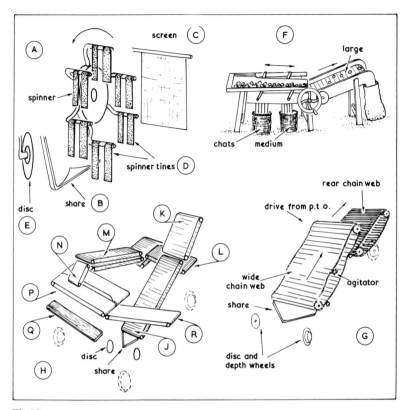

Fig 32

expose the crop properly. A canvas and net screen may be mounted on the machine to restrict the distance thrown (Fig 32C).

In other, lighter, soil conditions the tines are arranged to be at the same angle in relation to the ground at all points of their travel. This angle can be adjusted, but is near vertical. The effect is to give a push to the earth and crop, rather than a swing (Fig 32D). The mechanism to achieve this uses two wheels, with their centres staggered, driving the arms carrying the tines. A machine may carry a wheel like a plough disc-coulter that cuts ahead of the share (Fig 32E). Its main purpose is to cut away weeds and other trash that might otherwise get caught up in

143

the spinner. It also provides another step in loosening the soil for the spinner.

Potatoes, as they come from the field, have to be cleaned and sorted. Early attempts at improving on hand washing used an arrangement similar to that for other root crops. The potatoes were put into a large cylinder made of lengthwise slats, arranged horizontally and partly submerged in a trough of water. Turning the cylinder tumbled the potatoes against each other and the bars to wash off dirt. The trough was removed and a side door opened to let the potatoes fall onto a removable rack, where they were left to dry before bagging.

However, potatoes come in many sizes. The smallest (chats) have to be removed and used for animals rather than humans. If any of the potatoes have to be kept for seed, the medium ones are chosen. The obvious way to separate potatoes was to subject them to riddles or sieves with different-sized meshes. A commonly used type was often farm-made, and may still be found in use for comparatively small quantities. Known as a 'potato harp' in the north of Britain or a 'potato riddle' elsewhere, it had a wooden framework with two iron-meshed riddles, one above the other. The top meshes stopped the largest potatoes and let all others fall through. The second meshes let the chats through and stopped the rest. This gave three gradings and the potatoes were directed by chutes, or otherwise, into baskets.

This static riddle required labour to lift the potatoes before and after sorting. A further step was to use a cylindrical riddle set at an angle so that potatoes put in at the top were shaken down as the cylinder revolved. Small potatoes fell early through small spaces into a container, medium ones later through larger holes, while large potatoes went right through the cylinder.

Complete mechanization assumes that the article being treated is perfect. With potatoes there can be flaws that can only be recognised by hand sorting. Consequently a grader that allows human examination has advantages. Reciprocating riddles allow examination at all stages. In a typical sorter (Fig 32F), light enough to be hand operated, potatoes are tipped into a hopper which feeds them on to a high-sided frame with a riddle bottom. The mechanism gives this a swinging or reciprocating motion. The riddle has two sizes of holes, so chats fall through early into a chute leading to a basket or other container. Medium potatoes fall through lower down. The larger potatoes continue until they tip on to a conveyor that takes them to fall over the end into a sack. As the potatoes are spread over the conveyor, which is only travelling at a slow speed, it is possible to pick out faulty ones and remove them before the good potatoes tip into the sack.

As with the harvesting of cereal crops, the digging and preparation of potatoes has been united in a travelling, combined machine, although the first of its stages may be performed by a separate elevator digger. This is driven from, and towed by, a tractor. A share serves the same purpose as on a spinner machine, but earth, as well as potatoes, is gathered up on a conveyor (Fig 32G). The first stage of the conveyor is made up of an agitated web of bars, with gaps through which earth can fall. The potatoes drop from this on to a rear chain web, which then deposits them on the ground.

It seems illogical to lift potatoes then deposit them back on the earth again, so the next step is a complete harvester, which follows the elevator while travelling, and feeds the result to a trailer or lorry. There are several types, but basically the elevator digger is followed by conveyors (Fig 32H). Earth, stones and small potatoes fall through. Unwanted tops are brought up with the potatoes. At the top of the digger web (Fig 32J), an elevator with widely-spaced bars is positioned with sufficient clearance to miss potatoes, but take the tops (haulms) and drop them on the ground (Fig 32K). If any potatoes are still attached, they should drop through the wide spaces onto the transverse cross web conveyor (Fig 32L), joining those dropping directly from the digger web. This and subsequent conveyors may have rubber-covered bars to avoid damage to the potatoes, but there are gaps for dirt and stones to fall through at all stages. In some machines a caged-drum conveyor makes for compactness of the machine. The potatoes are delivered between parts of a double-web conveyor (Fig 32M). This may further press off clinging dirt. The potatoes and trash drop onto a small sloping web on rollers (Fig 32N), which is rotating with an uphill motion above two separated conveyor belts (Fig 32P). This sloping web acts as a final separator. Potatoes roll down to drop onto the front conveyor. Stones and dirt cling to it long enough to travel up and tip onto the rear conveyor.

It is at this point that mechanization has to give way to the human element, as men have to travel on a platform on the machine (Fig 32Q). Their main occupation is removing stones, and other things that have dropped onto the potato conveyor, and potatoes from the trash conveyor.

When these have been sorted out, the potatoes go on another conveyor (Fig 32R) to be deposited in a trailer, while the trash is returned to the ground.

Sugar beet harvesting

Sugar beet harvesting involves two main stages: topping and lifting. Several horse-drawn tools were devised for both jobs. The general construction was like a plough, but topping cutters, or lifting shares, took the place of the plough share. In a modern machine, while one row of beet is being lifted, the next row is being topped and the tops swept aside to keep them clear when that row is lifted next time along.

Topping is cutting off the green while only removing the minimum of the crown of the root. Cutting off too much crown is wasteful, while cutting too high, so that some green goes through with the beet, is unsatisfactory. This means that the topping mechanism must have many adjustments. In a common type the mechanism is driven from a land wheel. A feeler wheel brushes over the beet tops and a shaped knife below

Fig 33

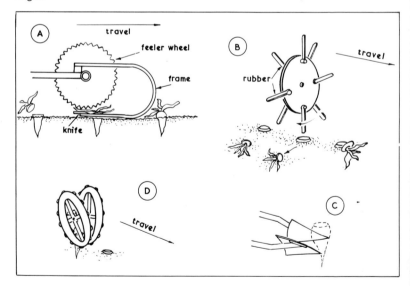

it slices off the top. Spring loading allows for uneven heights and a vary-ing thickness of top growth (Fig 33A).

The topping knife is followed by a flail, which is something like a potato spinner in its action. It consists of a disc with rubber blades (Fig 33B). Its height and speed are adjusted so that the tops are flicked to the side, out of the way of the digger. There may be a disc coulter running ahead of the topper to remove trash before it can impede the topping knife.

While the actual digging differs, the further stages are similar to those of an elevator potato digger. On some machines, lifting of the beet is done with a pair of shares (Fig 33C). As each beet is scooped out of the soil, it rides up the shares directly onto an elevator. Another method uses a pair of inward sloping Oppel wheels (Fig 33D) which squeeze the beet out of the ground. As each beet rises and the gap between the wheels gets wider, a flail working between the rear edges of the wheels pushes the beet on to the conveyor. Soil is shaken off the beet as it travels, by agita-tion, up the open-web elevator so that anything unwanted falls through. In some machines there is a fixed web above the elevator, the beet is rubbed against this and caused to move around and be scuffed. The beet falls from the top of this elevator onto a transverse one which delivers it to a trailer alongside, or else the machine may have a tank, from which the bulk is unloaded at intervals.

CHAPTER 9

Estate Management

Early farming did not have much use for fences, walls and hedges. Crops were strip farmed and animals grazed on common land. It was not until the time of the enclosures that much attention was given to fencing in Britain. Many boundary fences were, and still are, natural-grown hedges. Where stone is common there are walls, while wood and wire fences are comparatively modern.

Hedging and Fencing

A grown hedge has to be tended every few years if it is not to extend an inordinate amount, outwards as well as upwards. Traditionally, the hedger has used axe, slasher and billhook (Fig 19, p 95) to 'lay' a hedge. In this process, besides cutting back the hedge, parts of branches are slashed slightly and bent around posts driven in to make an impenetrable barrier, around which new growth will form again. This is still the best way, but it is time-consuming and much hedge trimming is now done mechanically, the more common type of cutter for farm-hedge maintenance today having a circular saw at the end of a long arm (Fig 34A). The saw angle can be adjusted from the tractor anywhere between horizontal and vertical.

Fence posts may be driven into holes dug with a spade or made with one of the special post-hole borers (Fig 19, p 95 and Plate 20). Today there are augers which work off a tractor to prepare the hole quickly and mechanically.

Wire used for fencing has to be stretched. One way of doing this is to put long-screwed eyebolts through the end posts and leave them in position. Several tools have been devised to tension wire before fixing the

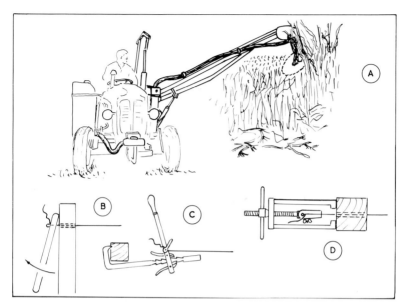

Fig 34

end. A simple lever could be used (Fig 34B). A refinement gave the lever two alternate fulcrums, via pawls working along a double rack (Fig 34C). A screw that pulled the wire with a plier-type grip was also used (Fig 34D).

Some rural craftsmen describe themselves as hedgers and ditchers. Many hedges have a ditch at one side which may carry a natural stream or be an aid to the drainage of the adjoining field. Mole drains (Fig 10E, p 62) must lead to somewhere so that the water can be carried away, and this is the purpose of a ditch. Ditching calls for special spades and scooplike tools (Fig 16, p 87) for removing mud and debris.

Forestry

Although the felling of trees and the conversion of timber does not now come within the sphere of farming generally, early farmers had to clear forest land and used timber for many jobs on the farm. Modern forestry equipment may not find a place on many farms, but most farmers are still able to make use of timber, which they convert themselves, even if the end product is simple fencing or just logs for firewood.

In their early development, axes probably followed a parallel course to hammers, mallets, picks and hunting weapons, being merely stone implements attached to shafts. With the coming of metals they became recognizable as separate tools and were probably the only ones available for felling and converting trees to man's use. Neither bronze nor iron will take much of an edge and there may have been flint axes better able to cut, but an axe severs wood only partially by cutting, as much of its effect is a splitting action due to its wedge section.

Some early axes had blades made by wrapping them around a shaft (Fig 35A). As smiths mastered the making of socketed heads, the tendency was to keep the socket as short as possible, for simplicity in construction, so axes developed with blades broader towards the cutting edge (Fig 35B). Many older, steel, axe blades that have survived have much deeper blades than are usual today (Fig 35C). Continental blades tended to sweep out to a wider cutting edge. Obviously, with axes made by local smiths, there were a great many variations. Like preferences for shapes of billhooks, some of these local prejudices survive. However, there was one characteristic of the majority of British axe heads: a broadening on each side of the socket (Fig 35D). In some parts the one-hand axe is called a hatchet, while the two-handed tool is an axe, but modern catalogues tend to call them axe and felling axe, respectively.

Axe hafts (shaft, handle) have to control the tool while transmitting the minimum shock to the user. In Britain, ash is the wood best able to do this. Hickory has similar characteristics and is the favoured alternative in America. Hafts have been many shapes, but the dog-leg pattern usual today could be seen in earlier axes, although some workers favoured straight hafts. The normal modern axe is a scientifically formed, all-steel, wedge shape, with its socket waisted so that the haft is secured by spreading with wedges—often a long wooden one, with a toothed steel one across (Fig 35E). The head has a uniform degree of tempering, but because of the simple dipping methods of tempering steel, used by the early smith, only the edge could be tempered, while the rest of the head remained soft. After the axe became worn and the edge was ground away, the new edge had to be re-tempered.

There were axes of many other forms, but these were mostly peculiar to country craftsmen (see the companion book *Country Craft Tools*).

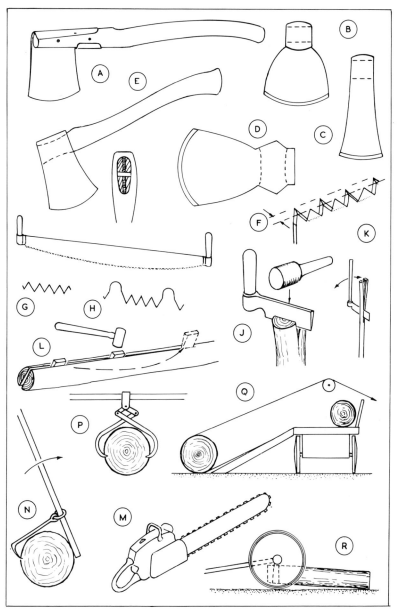

Fig 35

It was a long time before saws took their place alongside axes in forestry. Besides the difficulty of producing large, flat, saw-blades, saw-tooth design was not understood, and the uneven haphazard tooth form was even less successful on logs full of sap than it was on the drier wood of the workshop.

In cutting green wood, the saw has to make a kerf (cut) wider than the thickness of the blade if it is to cut through without binding. This is achieved by setting alternate teeth in opposite directions (Fig 35F). Many large crosscut saws had triangular teeth, filed to cut alternate sides of the kerf (Fig 35G). Having the teeth large helped in getting through damp wood and in clearing the sawdust, but providing gullets between groups of teeth (Fig 35H) clears the kerf better.

Crosscut saws for hand use are large and have handles at both ends. In use, each operator pulls in turn. This keeps the cut straight, while pushing might buckle the saw. Pit saws, of considerable length, with one man on top and another below, for cutting a log lengthwise into planks, were more the province of itinerant sawyers than farmers.

A technique long practised as an alternative to sawing along the grain, was splitting. This had an advantage in strength, as it followed the line of grain instead of breaking through it, but the resulting piece of wood could be far from straight, depending on the grain. Power sawing has almost killed this method, although it is still seen in manufactured chest-nut fencing. The tool used was a 'froe' (fromard, cleaving axe, framod, doll-axe, dill-axe). A froe has a straight cutting edge and a socketed handle (Fig 35J). It was started by a blow from a mallet or cudgel, then the cut was opened by levering on the handle (Fig 35K). Splits in larger logs were made with wedges. Ideally they were steel—wood could be used but had a limited life. Two or three wedges could be used in turn to drive in and make a split progress along a log (Fig 35L).

With the coming of steam power, the first mechanized saws were simi-lar to crosscut and pit saws held in frames and given a reciprocating motion. Some of these are still used, but the circular saw has taken over for general cutting, where the wood is brought to the machine. Most farms have a circular saw bench, either belt-driven or directly coupled to the power take-off of a tractor.

The modern tool which has taken over from both axe and saw is the self-powered chain saw (Fig 35M). The cutting teeth are on an endless chain powered by a small petrol engine, with the whole thing light enough to be lifted and managed by one man. There are versions powered by electricity via a cable, but these are obviously restricted to places within reach of power and usually for smaller jobs.

Bark was used in smoking bacon and in tanning leather. This could be stripped with an axe or one of the other cutting tools, but there were special spoon irons for the job. They varied, but were shaped like small rounded spades (Fig 18E, p 92) to get under the bark, with long shafts that were socketed to fit on the end of a pole.

Moving cut logs amongst trees and getting them out of the forest had to be done by man and horse power before steam engines became available. Besides the direct pull, use was made of levers. A 'cant hook' or 'log dog' had many forms, but basically it was a hook engaging with a pole to roll a log (Fig 35N). What is now called the 'lazy tongs' action, was used in a tool for lifting logs (Fig 35P).

Plate 29   A giant horse-drawn nib used for hauling logs in the redwood forests of southern Oregon, USA

Timber waggons, consisting of a pair of two-wheeled trolleys on a central stout member, and hauled by a team of four or more horses, have formed the subject of many pictures, but these were only suitable where there was a wide-enough track and access from the road. Logs had to be rolled up ramps. One method was to parbuckle (Fig 35Q), using horse power.

For less accessible places, where the weight was too great for a direct pull over rollers, a device on two high wheels was used, sometimes called a 'nib' (Fig 35R). As the log was hauled up under the axle, this had to be kept high by very large wheels so that the log would clear the ground. Only the forward end might be lifted; the other dragged. In countries where trees grow to a considerable size, as on the American west coast, wheels had to be very large (Plate 29), and in pioneer days had to be built up with available materials to give sufficient strength.

CHAPTER 10

Stock Management and Feeding

The shepherd probably has the longest tradition of stock management, and his equipment today is little changed from Biblical times. However, like billhooks, there are regional preferences for shepherds' crooks. These probably evolved from natural wood crooks—wood and horn have been used, but iron and steel have been normal ever since they became available.

The usual English crook has a curve forming about three-quarters of a circle, with a graceful sweep outwards at the end, finishing in a smooth knob or spiral, so as not to harm the sheep. This was of a size to catch a sheep by the leg (Fig 36A), and was on a long wooden handle.

Scottish and northern English shepherds favoured a crook large enough to catch a sheep by the neck. Some of these were made from wood, such as ash or hazel, bent while green and left to season to the fixed shape. A ram's horn could be softened by boiling and bent to this shape. As the shepherd's crook has been given ceremonial use in the church, it is necessary to be aware of the distinction between working crooks and these usually more ornate versions.

At one time sheep were treated with a salve as a protection during the winter, and to kill parasites. To do this meant working a long time on each sheep, during which the shepherd sat on the narrow end of a slatted stool, with the sheep on the broad part. In the nineteenth century this was superseded by dipping. To control the sheep passing through the dip a sort of double hook was devised (Fig 36B). One side could be used to push the sheep down, while the other end would pull the sheep's head above the surface.

Early shears for taking wool from a sheep looked like large scissors with blunt ends (Fig 36C). Factory production in the nineteenth century

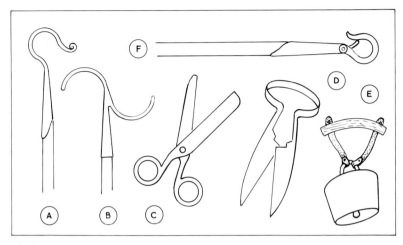

Fig 36

resulted in the type of shears with a sprung arc at the top (Fig 36D). Although these have given way to mechanical shearing, they are still commonly available, used for trimming fleeces and other jobs.

For mechanical shearing, the tool at the end of a flexible shaft or cable works like barber's clippers at a very high speed, with a cutter bar crossing a fixed bar to give a scissors action. Power may be provided by electricity or the power take-off of a tractor, although there were tools of the same type driven by a hand crank.

As with other animals allowed to roam in search of food, sheep were sometimes fitted with bells. Wooden yokes were used (Fig 36E), but straps were more usual. Sheep bells were often narrower at the mouth than at the top, and many were very crudely made of wrapped sheet iron. Some iron bells were brass-plated to improve their tone. Clappers were iron or bone.

Goatherds and swineherds had little equipment and what they had was often based on what the shepherd used. Cows that roamed on common land were given bells that differed from sheep bells only in being larger. If a bull had a ring in its nose, it was handled with a stout pole, having a clip in the end to grip the ring. This bull leader (Fig 36F) needed a secure spring closure and some were arranged to be released remotely by a cord. If the bull had no ring, two rounded jaws could be sprung or pinched into its nostrils. These devices are still used.

Branding irons were used on cattle, with a foot, forming the device of the owner, heated in a fire and used on the end of a handle. In Britain the device was more likely to be initials than the more colourful ones of the American cattle ranches. Branding is going the way of the cowboy's lariat, as indelible colours can now be used.

Veterinary work

The range of veterinary tools is extensive and the number used by a farmer himself would have depended on his skill and whether he was out of reach of a skilled practitioner, so the number that can be regarded as his tools is variable. Various types of tools with a shearing action were used for docking tails and other operations (Fig 39H, p 167), although for lambs a knife might be all that was needed. Forceps for use in lambing were patented at the end of the nineteenth century (Fig 39J). Blacksmith-made gags for horses (Fig 39K) were used to hold their mouths open, either to inspect teeth or administer medicine. Pellets of medicine were blown down the horse's throat with a blowpipe, sometimes with a piston to provide the pressure (Fig 39L, p 167).

Food preparation

Until comparatively recent years, most farms depended on what they could grow to feed livestock and poultry and any necessary preparation was done on the farm. Most early barn machines were concerned with the preparation of animal foods and many of these would certainly not come up to the safety standards of a modern factory inspector.

There were many hand tools. Turnips for feeding sheep and other animals needed to have leaves removed and be chopped up. The knife for the first stage was like a billhook with a spike instead of the curved end (Fig 37A). Broken or worn-out sickle blades were sometimes converted to this job, so the knife part was not necessarily straight. The spike picked up the turnip, so the worker did not have to bend so far. To cut the turnip into pieces it was pounded with a tool fitted with crossed blades. This might have worked end-on with a straight handle (Fig 37B), or be at an angle to the handle so that it was used like a hoe (Fig 37C). A similar tool with two parallel blades was used to cut turnips into slices.

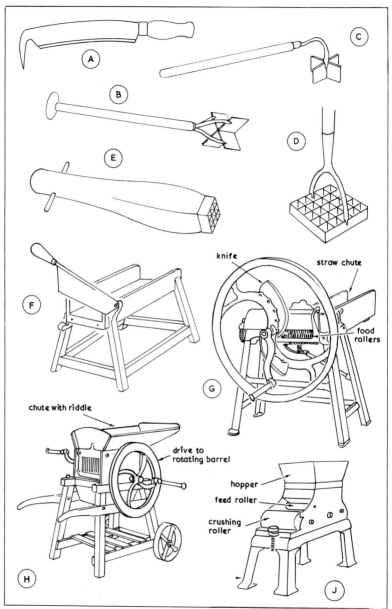

Fig 37

158

A tool of similar appearance was the 'barley hummeller'. This had a greater number of blades, which might be parallel or arranged to cross (Fig 37D). This was pounded on the barley to remove the long spines or 'awns'. A roller type, which gave the same action, was also used and later threshing machines incorporated hummellers to deal with barley. A mechanical hummeller has a sloping perforated cylinder, into which the barley is fed. Inside, iron beaters throw it around so that the awns are completely broken away.

Another tool with an end like a barley hummeller is a 'whin bruiser'. This differs in being made of quite a bulky piece of timber to provide weight (Fig 37E). Whin is another name for gorse, more commonly used in Scotland. To make gorse shoots into animal feed they have to be pounded to a pulp, hence the heavy weight of the bruiser. This involved strenuous work for a limited output. Where gorse was plentiful and the need greater, a mill, similar in principle to that used for crushing apples for cider, was used. A stone wheel was drawn round a post by a horse. Gorse was forked under it and turned over between revolutions.

Straw had to be cut into short pieces for feeding animals, and horses, especially, required quite a large quantity. The first stage, from cutting by hand with a knife, was a simple cutting box. Straw was pushed along a trough and a long hinged knife was pulled across the end (Fig 37F). In a better box, the end of the trough had a lower cutter to produce a scissor action.

This simple tool developed first by the addition of foot-operated holding arrangements and hand pushers, until many inventors produced their own versions that became increasingly automatic. Many of these later machines used a large flywheel with two rotating knives and the straw, or chaff, feed became automatic (Plate 30). In a typical machine (Fig 37G) the uncut hay or straw is put into a chute and gripped by a toothed roller that feeds it towards the rotating knives. Enclosed gearing controls the amount of feed and therefore the lengths of the cut pieces, and these gears can be changed. In some makes the worm wheel, which transmits the drive from the hand wheel, can be turned round to allow two speeds of feed, giving longer pieces for cattle and horses, or very short pieces for sheep.

Hand-driven machines of this type were in use near the end of the nineteenth century and many of them are still in existence and being worked today. Other methods of cutting were tried. Some machines

passed the straw between two rollers, one of which carried sharp knives which pressed against the other, but the shearing action of knives on a flywheel proved more successful. As steam power developed, larger machines came into use and the output was able to keep pace with the needs of users of large numbers of horses.

Turnips, swedes and other root crops can be processed by machine much quicker than by hand, before being fed to cattle and sheep. The hand-driven machine, used for a long time, operated in a very similar way to a kitchen mincing machine, and in the same way could have different cutters to produce different effects. Roots needed cleaning first. In one type of cleaner there is a tilted cage made up of many bars, some of which are bent to throw the roots about as the cage rotates and they are shaken towards the lower end, where dirt falls out between the bars. A refinement allows the angle of the cage to be altered and dirtier roots can take longer if the angle is made shallower. This machine works with reasonably dry roots, but some machines had the cage turning in a water trough to wash off mud.

In a basic root cutter, driven by handles at opposite sides, the roots are fed by a hopper into a rotating barrel with knives on it. The cut pieces fall through into a container, or leave by a chute. In another type of machine, the roots were sliced by a knife operating rather like a chaff cutter, with a choice of two thicknesses of slice by changing the direction of turning the handle. Many hand machines were made portable so they could be moved on wheels by a pair of handles and taken to the field as well as being of use in the barn (Fig 37H).

Although some animals eat whole grain, it is more usual to crush it by milling first, to help mastication. For grinding on the farm, instead of using the local water- or windmill there have been several types of machine to prepare grain for cattle. Some of those date from the eighteenth century. Cracking maize for poultry is called kibbling. Oats for horses are described as rolled. Corn crushers do this work with rollers of the mangle type, which can be adjusted according to the amount of crushing needed. For oats the rollers are smooth; for maize they are

Plate 30   This chaff-cutter has ended its days as a sign at the entrance to Haddenham museum

fluted. In a typical machine, corn goes into the hopper and is directed by a feed roller to the crushing rollers, then falls through the bottom of the machine (Fig 37J).

To grind the meal finer a plate mill is used that works on the same principle as the large mill, but instead of stones there are vertical chilled-iron plates. One of these discs is stationary and the other revolves close to it. There are grooves in the discs and the grain is fed between them near the centre, usually by a worm or auger device. The grooves are shallower near the centre and the discs made so their surfaces are closer towards the rims. The meal is broken finer as it works its way towards the edges of the discs. Adjustments are provided to suit different grains and to get the degree of fineness required. The grain has to be riddled as it is fed to the plates to remove stones or other matter that might damage them (Fig 38A). Some early mills of this type used buhrstone instead of chilled iron and, even with iron plates, the type of mill was sometimes called a buhr mill.

Another method of producing meal is to use a hammer mill. In this a large number of hammers are turned at a high speed by a rotor. The hammer faces pound the grain above a screen until it is small enough to fall through. Screens can be changed to regulate the size of the particles of meal. The action of a hammer mill is such that it will grind almost anything to meal. This means that unthreshed corn can be put through and the straw will be pounded to meal with the grain, which is an advantage when roughage is wanted with the flour.

A hammer mill has to be power-driven, either electrically or by a tractor. The hammer action requires a simple fast rotation. Grain is fed to the hammer chamber through a hopper. The flour may fall into a trough, and a fan on the same shaft as the hammer rotor blows it, via ducts, to a reservoir, from which it can be fed by gravity into bags (Fig 38B).

Oilcake was one of the early bought-in cattle foods. As supplied, oilcake is extremely hard and in large pieces which have to be broken up, so, when oilcake came into use from around the turn of the nineteenth century, farms had to instal cake-breaking machines. The majority of these used pairs of rollers of the mangle type, but with spikes, grooves or ridges to break the cake. In a typical machine the oilcake was dropped by hand into a box or hopper so that it fell between spiked rollers. Power from the hand wheel was geared down to increase strength. The effect was to pro-

162

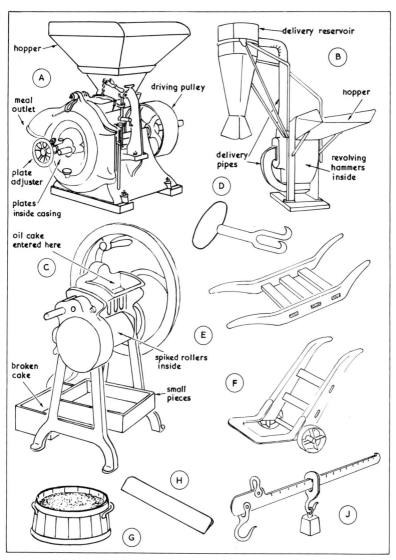

Fig 38

163

duce a certain amount of powder or dust as well as pieces of manageable size. The powder fell through a grating into one container, while the pieces were led by a chute to another (Fig 38C). The broken pieces were fed to cattle, while the powder was fed to sheep or used directly on the land as manure. Later power-driven machines had another pair of spiked rollers, with both pairs adjustable, so the size of the final pieces could be regulated. More recently, oilcake and other concentrated feeds have become available as biscuits or pellets of suitable size, without the need for breaking.

Before the days of plastic bags, much cattle food had to be handled in hessian sacks, mostly of a size bigger than a man could manage easily. Sack hooks (Fig 38D) helped in lifting these. A sort of two-man stretcher was also used for carrying sacks (Fig 38E), and sack trucks or barrows (Fig 38F) evolved as a means of easing a sack off the ground and wheeling it with the minimum labour.

Much use was made of baskets. As a container, a basket is lighter than any other rigid box and has the advantage of being made from materials obtainable locally. A fairly open weave would be suitable for root crops, but others were made with a closeness of weave and construction suitable for grain. Wrapped strips laid around could be made tight enough for almost any content. Baskets were used to move roots, grain, seed and similar things around the barn or to the fields. Pails were made of wood staves, like barrels, but later both these and baskets became replaced by galvanized iron buckets and baths which, in turn, have given way to plastic goods.

There were usually several sieves kept near the food-preparing machines for hand sifting. These had meshes of different sizes formed with wire or wood strips and were framed with thin wood, bent around the outsides. Various wooden hand scrapers were devised for moving and stirring meal.

Most farms had some means of measuring, either for use when compounding feedstuffs or when buying or selling. The bushel was the usual standard of quantity and the usual means of measuring was a sort of broad shallow pan, made with staves like a barrel (Fig 38G). The contents were levelled across the top by passing a roller, or sliding a flat-edged board, across. In both cases the tool was called a 'strike' (Fig 38H). The commonest means of weighing was the 'steelyard', which goes back into antiquity and has the advantage of being able to weigh a considerable amount without the need for a large number of heavy weights (Fig 38J). They were not always steel and sizes varied considerably. The more permanently installed types had platforms to take sacks and other bulky items.

Of course national standards of measurement were not enforced and there could be local variations, particularly as the lord of the manor was all-powerful. The rod, pole or perch, which settled at $5\frac{1}{2}$yd (16ft 6in) before it was discarded, was quoted on one manor, as 18ft 6in of the landlord's land and 18ft 0in of the tenant's. Bushels varied according to the grain. Until quite recent days a bushel of oats was a different quantity from a bushel of barley.

Besides brewing on the larger farms, cidermaking was also common on small farms, particularly in the West Country and East Anglia. The apples used have to be crushed to a pulp then pressed to extract the juice. What is left can be fed to pigs.

Crushing was done in crude mills, with the apples fed past spiked rollers. Better machines then took the broken apples between smooth rollers. On a bigger scale, a large stone wheel (discarded mill wheel on edge) was made to revolve in a round trough by a horse walking around at the end of a beam pivotted at the centre of the round trough. It was considered advantageous to crush the whole apple, including its pips, to get the best flavour, so powerful crushing was best. On a smaller scale, pulp was made by jumping a pole or rammer on the apples in a trough.

The idea of the cider press is very basic, being merely a heavy weight to force the juice out into a channel which led it to a container. A simple press consisted of a perforated box with a sliding lid and a grooved channel around an extended baseboard. In use, the pulp (pomace), between layers of straw, was put in the box and pressure applied to the lid. The pressure might come from direct weights, although their effect was more likely to be increased with levers. Better presses had large screws to apply pressure. Complete extraction of the juice took many days. Apple mills and cider presses still survive, but commercial cider production is now a much more hygienic, factory operation.

CHAPTER 11

Dairy Produce

Milking

Man has milked animals since very early days and the first were more likely goats and sheep than cows, but it is probably that the greatest changes have taken place in milking cows.

For hand milking the operator sat on a three-legged stool, with his or her head pushed into the cow's side, and manipulated her teats to direct the flow of milk into a bucket. Equipment was simple, the technique straightforward, but the work output was slow.

There were experiments in mechanical milking from about the middle of the nineteenth century. In 1862 an American, named Colvin, produced a device with teat cups connected to a bucket and levers to produce a vacuum. The idea of a vacuum was right, but the device lacked the necessary pulsating action of calves sucking. Other experimenters tried to imitate the hand action.

The first milking machine with any degree of real success was produced by William Murchland of Kilmarnock, in 1889, who used a vacuum controlled by a column of water and continuous suction. This was a permanent installation in a shed, with air drawn from a pipe by a hand pump. Stopcocks were on branch pipes to each stall, so several cows could be milked by the action of one pump. The vacuum was at the milk bucket and milk was sucked down through the teatcups. As with earlier continuous vacuum machines, the process was painful for the cow and damaging to her udder.

A machine produced by Dr Alexander Shields of Glasgow, in 1895, imitated calves' feeding action. It was an expensive piece of equipment, but it paved the way for modern machines.

Although the principle used may be the same, several methods of assembling the cows are used. On a modern farm there may be a cowshed large enough for the whole herd at one time. Here the cows are in rows, tethered by the neck, with a manger of food in front of them, and a pipe-line from the milking machine is led above them. An alternative is a milking parlour. These take only a few cows at a time. The milking parlour is likely to be more modern in layout and feeding may be automated. A milking bail is a mobile or portable milking parlour. This is powered by a tractor and taken to the cows when they are on summer pasture distant from the farm buildings.

In a modern milking system there is a vacuum pump (Fig 39A) driven at a constant speed by an engine or electric motor. Along the permanently installed line is a sanitary trap (Fig 39B), which is a filter to stop moisture or dirt being sucked into the pump. Although the pump will normally maintain a steady vacuum, there is a control valve (Fig 39C) which automatically keeps the same degree of vacuum. A gauge (Fig 39D) registers the amount of vacuum, which is maintained at about half atmospheric pressure. Branch pipes (Fig 39E) have stall cocks (Fig 39F). How the milk is dealt with depends on the type of installation, but in a

Fig 39

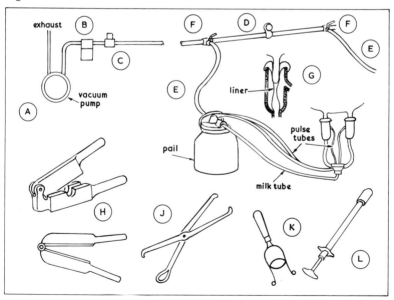

bucket system there is a large pail with an assembly at the top. This includes a device in which air is pulsated to enter and then draw out again from a space between the teat cup shell and its liner, so that the liner alternately squeezes and relaxes (Fig 39G), thus drawing milk. The machine is adjusted to give forty to sixty-five pulsations per minute. As it is when the liner is released that milk is drawn, this is usually arranged to be slightly longer than the squeeze period, so milking is performed quicker. However, anything approaching the direct uninterrupted vacuum of some early machines has to be avoided, as this is harmful to the cow.

Milk, with butter and cheese produced from it, was more a family or community occupation than a business, until well into medieval times. In a rural economy each family could supply its own needs from one or more cows.

Obviously a multiplicity of small units, using simple equipment to produce small amounts of anything, is not the best economic method, but where people were self-sufficient, as far as possible, and trade of any sort was quite a small part of their lives, the method worked and seemed acceptable. Changes did not come until townspeople engaged in factory work had to abandon the idea of getting milk and its products from their own cow, and began to buy or barter for them. This led to dairy herds becoming larger, and the farms developed dairies of a more advanced type, capable of handling larger quantities of milk. Today most milk leaves the dairy farm to be processed in a factory and even the farmer who owns the herd may be buying his own milk from the factory in a bottle.

Milk and its products, as an industry, date from the seventeenth century, when sales from the country to towns began to grow. The dairy industry had to remain a fairly local enterprise for a long time because of the short keeping life of milk, but this was altered with the introduction of refrigeration, which came first in America in the 1850s. By the beginning of the twentieth century the dairy industry away from the farm was well established.

Most dairy implements were of wood, and light-coloured hardwoods were favoured because of their clean appearance. Maple and sycamore are near-white woods that were much used. Beech was darker but was sometimes used for carved or more intricately worked objects. Pails, churns, mixing vats, presses, and anything round were made in the same

Fig 40

way as barrels, with closely fitting staves and iron bands around. The capacity of a milk pail (kit) was about six gallons. As five gallons is as much
as most men are able to carry far by hand, and milk was more likely to be
carried by women, a yoke to fit the shoulders was an essential part of
dairy equipment. This was carved from one piece, and hooks on chains
used to balance two pails (Fig 40A). Some pails had the handle common
to modern buckets, or one stave extended to make a handle (Fig 40B).

Much milk was used straight from the cow at about 90°F (30°C), without treatment. If the milk was to keep any length of time it had to be
cooled to about 50°F (10°C). A small quantity could be merely left in a
cool place, but on a larger farm the milk passed through a cooler, in
which it ran over corrugated metal or pipes with water passing through
to lower the temperature. This type of cooler, which is still used, was
invented at the middle of the nineteenth century. It needed piped water,
but that could be a local supply. Milk had to be poured in by hand at the
top and collected in a churn at the bottom.

Milk churns, tapered to the top for stability, provided the means of
transporting milk from the farm, either to a factory or a distributor.
Retailers delivered milk from a churn of about ten gallons' capacity (Fig
40C) from a horse-drawn cart or a hand barrow. Using measures to pour
the milk into the customers' jugs continued in Britain at least until the
1920s when bottled milk took over, bottles having been customary in
France as early as 1885.

The only other farm treatment for milk was straining, to remove any
solid matter. This might have been done with a sieve at the cooler, or a
hand sieve or straining cloth. Today, of course, there are further steps in
the factory treatment of milk, in particular pasteurisation, named after
Louis Pasteur, who discovered this means of destroying harmful bacteria
in cows' milk, and the process became usual from the beginning of the
twentieth century.

Cream is removed from milk to make butter. If milk was left up to a
day in a broad shallow pan, the cream would come to the top, and the
separated cream could then be removed with a skimmer (Fig 40D), in the
form of a broad, shallow, handled, pan, usually metal, but occasionally
of wood. Some small holes at the centre allowed any milk to drip
through, but the cream was too dense to leak out. The cream from one
milking was probably not enough to justify making into butter, so it was
stored. A stone or earthenware jar was favoured as this kept the cream

cool, but metal cans were also used for storage purposes.

Separating pans or dishes for a large quantity of milk occupied a large part of a dairy for a long time. A larger quantity of milk could be put in a deep, sealed, container, with a little air space at the top and submerged in water. This was found to produce good cream for butter-making and the exposure of the surface of the milk to the atmosphere was avoided. In the simplest form the water was static, but later models used flowing water.

A Swede, Dr Gustave De Laval, invented a successful milk separator with the centrifugal method, in 1878. The following year the machine came to England and proved itself by taking skimmed milk from which cream had been separated by other methods, and getting more cream from it. The De Laval machine was speedy and had little bulk in relation to its output (Fig 40E). Developments from it are used in the modern milk industry.

In a centrifugal separator, milk is heated and fed, via a float valve, into the top of the machine, where it goes through to the centre of a container spinning at high speed. The heavier milk flies outwards and the lighter cream stays at the centre. The skim milk rises around the side of the container and finds its way out. The cream rising up the centre reaches a different container and another outlet. The important thing is the high speed of the spinning container. At the time De Laval invented the separator, steam power was not generally available, so the power of a horse, walking round a vertical shaft, had to be geared up through a chain of belt drives or gear wheels to give sufficient speed. Later machines were powered by internal-combustion engines or electricity, although De Laval, and those who copied him, produced machines that could be hand-operated for small quantities as well as other sizes, up to those needed in factories.

Butter-making

Butter is made by agitating cream hard enough and long enough. It can be produced with enough patience by shaking by hand in a jar, but butter churns are the most satisfactory devices for reducing the labour and speeding the process.

One of the earliest types was a 'plunger' churn (Fig 40F), made after the manner of a barrel, but with the bottom larger than the top. The lid could be tightly fitted and fastened in place after the cream was poured in. Through a hole at the centre of the lid there was the handle of a plunger, which was a circular block with holes in. Butter was made by moving the plunger rapidly up and down. The plunger churn goes back at least to the fifteenth century. Later there were mechanized versions, driven by steam engines through a rocking crank, some with more than one plunger, but 'barrel' churns were more adaptable to this purpose. Even for hand operation, a crank handle was easier to use than a plunger.

In a barrel churn of the simplest form, the barrel was supported on an axle through bearings on a substantial trestle framework, with a handle at one or both ends (Fig 40G). Baffles or diaphragms inside were perforated and had the effect of stirring the cream as it was tossed from end to end with the turning of the barrel. Refinements were in the method of opening and sealing, as it was obviously necessary for there to be easy access, both for removing butter and cleaning the interior.

There were also barrel churns with the axle through their length, which stirred the cream by being of eccentric cross-section. Any method of agitating would make butter, so there were a great many ideas tried, including rotating paddles. A glass jar type with a crank handle to turn the paddles (Fig 40H) is still made in America for use by those who have a nostalgia for the old days. Another method used a rocking box with baffles to give a stirring effect.

The alternative to the barrel for small farms was the 'box' churn, in which radial paddles did the stirring. The paddles took various forms, usually arranged four in a cross, either perforated solid blades (Fig 40J) or built up like ladders.

After the butter had formed there would be water and buttermilk to drain away. The butter was washed with clean water and with a salt solution—this gave it the familiar taste and a longer life. However, the butter still contained an excess of water and whey, which had to be removed. This could be done by kneading in the hands, but it was more effectively done with a butterworker. There were hand and mechanical butterworkers, but the action was to squeeze between a fluted roller and a flat platform, with guides on which the sides of the roller ran (Fig 40K). Working had to be continued until nothing more ran out.

At this stage the butter might go in bulk to be divided and made into pats elsewhere. If it was to be used on the farm or sold direct from it, it was worked into suitable blocks with butter pats or 'scotch hands', which were wooden spadelike tools, used in pairs (Fig 40L). Normally grooves were on one side and the other side was plain. There were also butter moulds, carved with designs to press into butter to provide decoration (Fig 40M).

Cheese-making

Cheese is mentioned by Biblical and ancient Greek writers. The Romans, who merely allowed milk to sour slowly, were great cheese-makers and may have introduced cheese-making to Britain. Cheese is normally started by adding rennet to milk to sour it. Rennet can be pre-pared chemically, but was usually in the form of a calf's stomach in a muslin bag.

There are a large number of cheeses with peculiarities about their making, but basically the milk is stirred and the rennet added. Curd forms into lumps, which are gathered, and what is left is whey. Curds gathered together will make the cheese.

The curds, as they formed, were stirred by a sort of open-mesh fork, called a curd agitator. Before being pressed into the cheese block the curd had to be repeatedly cut. Besides a knife there were cutters, some-thing like the turnip cutter (Fig 41A), but to resist the acid of the rennet these were protected by tin plating.

When the cheese was ready for pressing it was put into a mould (cheese vat, chessart). This was a very substantially-made parallel-sided barrel construction (Fig 41B), without a top and usually without a bottom. If there was a bottom it was perforated to allow whey to escape. Later chessarts were iron.

Fig 41

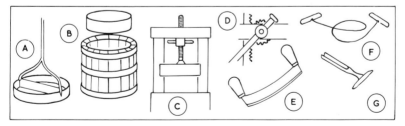

For pressing, a thick lid (sinker) was arranged to slide inside the chessart. It was thick enough to project above the top and take the pressure from the press. Cheese presses had a family likeness to cider presses. Stones might have provided the weight, but by the nineteenth century they were given a screw or lever action. The screw was a simple, central, threaded rod in a substantial framework (Fig 41C). The lever type probably took something from a type of press then used for printing. The lever acted through a ratchet (Fig 41D). In any type pressure was progressively increased. After pressing cheeses had to stand for some time, being turned over every day.

Cheese could be cut with a knife, although a two-handed type was needed on some (Fig 41E). Another way of cutting was to pull through a wire (Fig 41F). The quality of a cheese was tested by tasting, and for this a gouge-like tool was pressed in and turned to remove a finger of cheese for sampling (Fig 41G).

Sources of Information

There are, of course, a very large number of firms engaged in agricultural engineering and a great many museums which feature agricultural implements. Addresses given here form a very abbreviated list and there could have been many more, but those mentioned are those known to the author. Anyone wishing to learn more will find much of interest at any county agricultural show, while the Royal Show at Stoneleigh in the summer, and the Royal Smithfield Show in London in the winter, with their catalogues, are the best guides to modern agricultural implements.

The Royal Agricultural Society of England, National Agricultural Centre, Stoneleigh, Kenilworth, Warwickshire, is the leading authority in the country. The National Federation of Young Farmers Clubs, YFG Centre, at the same address, organizes practical activities in many parts of the country. The Council for Small Industries in Rural Areas, 35 Camp Road, Wimbledon Common, London SW19 4UP, can provide details of rural craftsmen.

Museums and collections

Up to date information on these and other museums can be found in the annual *Museums and Galleries in Great Britain and Ireland*. It is published in July by ABC Travel Guides Ltd, London Road, Dunstable LU6 3EB.

While the larger and more general museums are open on most days, the more specialized collections may be subject to limited opening times and these should be checked before visiting them. Where a figure number is given in brackets, this indicates that the implement sketched is housed in the museum.

Angus Folk Museum, Kirk Wynd, Glamis, Angus, Scotland
Avoncraft Museum of Buildings, Redditch Road, Bromsgrove, Worcester
Bicton Countryside Museum, Bicton Gardens, East Budleigh, Devon
Farmland Museum, 50 High Street, Haddenham, Cambridgeshire
Guernsey Island Museum (8)
Jackson Collection, Old Kiln Museum, Reeds Road, Tilford, Farnham, Surrey
    (7E)

Mary Arden's House, Wilmcote, near Stratford-on-Avon, Warwickshire

Museum of English Rural Life, University of Reading, Whiteknights, Reading RG6 2AG

Oxford City and County Museum, Woodstock, Oxfordshire

Rutland Museum, Catmos Street, Oakham, Leicestershire

Science Museum, South Kensington, London SW7 2DD

Weald and Downland Open-air Museum, Singleton, Chichester, Sussex

West Yorkshire Folk Museum, Shibden Hall, Shibden Park, Halifax, Yorkshire

White House Stedman Homestead, Aston Munslow, near Craven Arms, Shropshire

Cades Cove and Pioneer Farmstead, Great Smoky Mountains National Park, North Carolina/Tennessee, USA

Museum of Science and Industry, Chicago, Illinois, USA

Smithsonian Institution, The National Museum of History and Technology, Washington DC, USA

Agricultural implement manufacturers

These few firms have experience going back into the history of agricultural implements. There are many more. The names of some of them can be found in the advertisement pages of any farming magazine.

British Leyland, Tractor Operations, Bathgate, West Lothian EH48 2EF, Scotland

David Brown Tractors Ltd, Meltham Mills, Huddersfield, Yorkshire HD7 3AR

Ford Motor Co Ltd, Tractor Operations, Cranes Farm Road, Basildon, Essex

International Harvester Company of Great Britain Ltd, 259 City Road, London EC1P 1AD

    401 N Michigan Avenue, Chicago, Illinois, USA

John Deere, Harby Road, Langar, Nottingham

    Moline, Illinois, USA

Massey-Ferguson (United Kingdom) Ltd, Banner Lane, Coventry CV4 9GF

New Holland Division, Sperry Rand Ltd, Gate House Road, Aylesbury, Buckinghamshire

    500 Diller Avenue, New Holland, PA17557, USA

Ransomes, Sims & Jefferies Ltd, Nacton Works, Ipswich, Suffolk IP3 9QG

Glossary

Auger   tool for boring holes, either by hand in wood or tractor-driven
Awn   the beard of barley

Badikins   *see* whippletrees
Bagging hook   large version of sickle
Balance plough   *see* plough
Baler   machine for compressing and tying hay (*see* pick-up baler)
Barley hummeller   tool for pounding barley
Beam   main structural part of plough
Beetle   large mallet
Billhook   short-handled swinging knife
Blacksmith   worker in hot iron
Breast   *see* mouldboard
Breast plough   pushing, spade-like tool for skimming top of soil
Breast wheel   water mill with water led opposite the middle of the wheel
Bridle   *see* hake
Broadcast sowing   scattering seed, usually by hand
Broad share   ploughshare that pares without turning over
Broom squire   maker of besom brooms
Buck rake   tractor-mounted hay sweep
Bull leader   hooked pole to engage ring in bull's nose

Cambridge roller   roller built up of a large number of rings
Cant hook   *see* log dog
Cart   two-wheeled horse-drawn vehicle
Cas chrom   foot plough, lever-action digger
Caterpiller   *see* tractors
Cavings   chaff, broken ears and waste from threshing corn
Chain saw   portable powered saw
Chat   very small potato
Chat potatoes   smallest potatoes
Cheese taster   tool for removing sample of cheese

Chessart   cheese vat for pressing
Chisel plough   *see* plough
Clevis   *see* hake
Cock   loose-cut crop stacked in field to dry
Compression ignition engine   internal-combustion engine where the fuel mixture ignites under compression instead of with a spark
Cooper   maker of barrels
Coulter   knife or disc to split soil ahead of plough share (colter—USA)
Cradle   attachment to scythe to gather grain
Crawler   *see* tractors
Culls   animals which are discarded
Cultivator   tool for opening soil, working deeper than a harrow

Dibbing (dibbling, setting)   planting seed in holes made with a dibber
Disc plough   rotating discs used instead of share and mouldboard
Dogstick   dragging strut to prevent waggon running back
Drill, seed   *see* seed drill

Electric fence   electrified fence to give shock to animals
External-combustion engine   where fuel is heated outside the engine—as in a steam engine

Fallow   land left bare
Fantail   auxiliary sails on a wind mill for turning the main sails into the wind
Farrier   maker and fitter of horseshoes. Usually also blacksmith
Fiddle, seed   carried mechanical device for broadcast sowing seed
Flail   hand tool with swinging beater for threshing corn, or swinging arms on spinner in harvesting machine
Fold   movable pen
Foot plough   horse-drawn without wheels, but with skid at front
   see also cas chrom
Forage harvester   machine to cut and shred grass for silage
Furrow   the space left when the plough has turned over the furrow slice
Furrow press   roller to make furrows for seed

Gasolene   alternative name for petrol, more usual in America
Gate hurdle   hurdle made from strips nailed together in the form of a gate
Gavel   hook for pulling cut grain together
Grapple   bent fork for dragging (scratter, spike hoe)

Hades   grass strip between strips in open field
Hake   attachment on plough for pull by animal (muzzle, bridle, clevis—USA)
Hales   plough handles
Hammer mill   mill working with pounding action
Harrow   tool for loosening and dividing surface of soil

Haulm   the stalks of green crops, particularly potatoes
Hay   grass cut and dried
Hay bond twister   tool for making straw rope (wimble, throw hook)
Haymaking   drying grass and turning it into hay
Hay sweep   pronged scoop for collecting hay
Hoe   hand or machine tool used mostly to control weeds
Hopper   tank or chute to hold grain etc
Horse power   unit for measuring rate of doing work
Hummel   to knock off awns of barley
Hummeller   *see* barley hummeller

Internal-combustion engine   where power is provided by burning fuel inside
   the cylinder

Kerosene   alternative name for paraffin, more usual in America
Kibble   crush or bruise corn or beans for cattle feed

Laid corn   the crop does not stand upright
Landside   edge of unploughed land
Levelling box   *see* mouldbaert
Log dog (cant hook)   lever for rolling logs

Machete   swinging knife, similar use to billhook
Mattock   single or double-bladed chopping tool
Mill   any device for grinding meal and corn
Mill race   water above a water wheel
Mole plough   device for making drain holes
Mouldbaert   levelling box used to move soil from high to low spots
Mouldboard (breast, wing, turn-furrow, moldboard—USA)   part of plough
   behind the share which turns over the furrow
Mowing   cutting grass
Mowing grass   standing grass awaiting cutting
Muck spreader   trailer for distributing manure
Muzzle   *see* hake

Nib   two-wheeled cart for lifting and dragging logs

One-way plough   *see* plough
Osier   willow (withie) for basketmaking
Overshot water wheel   water mill with water led to top of wheel

Paraffin   vapourizing oil less volatile than petrol (kerosene—USA)
Petrol   volatile fuel used in internal-combustion engines (gasolene—USA)
Pick-up baler   machine to gather and bale hay in the field
Pitchfork   fork, usually two-pronged for moving hay

179

Plate mill   mill for grinding meal between metal plates
Plough (plow—USA)
    balance.   Type of reversing plough with parts repeated at opposite ends
    breast   *see* breast plough
    chisel   tool with deep-cutting tines to break up soil
    disc   *see* disc plough
    foot   plough without wheels, but with skid
    one-way   plough turning furrow always to the same side
    paring   used to skim off turf
    reversing   plough cutting either way, usually with duplicated parts
    ridging   used to raise ridges, as earthing up potatoes
    riding   plough on which driver had seat, sometimes three-wheeled
    stump-jump   spring-release share to clear obstructions
    swing   plough without wheels or skid
    turnover   type of reversing plough with parts arranged to rotate on a shaft
    turnwrest   plough that can be altered to turn furrow either side
Ploughing (plowing—USA)   turning up land by cutting furrows
    cable   drawing plough by cable, particularly with steam engine
Pomace   pulped apples for cider
Post mill   a wind mill where most of the mill can turn on a central post
Potato spinner   machine to dig potatoes with spinning tines
Prairie breaker   large plough for initially breaking new land
PTO   power take-off, an auxiliary drive from a tractor

Riddle   a type of sieve
Ridged roller   roller for forming ridges
Riding plough   *see* plough
Riff (ripe stick, strickle)   wooden sharpening block
Reaping hook   large sickle
Reversing plough   *see* plough
Roller crusher   machine for bruising cut grass

Sack hook   handled hook for lifting sacks
Scotch   roller behind wheel to prevent waggon running back
Scutch   remove dry husk of flax
Scythe   two-handed swinging knife
Seed drill   device for planting seed in rows
Seed fiddle   *see* fiddle
Seed-lip   basket or holder from which seed is sown broadcast
Share   pointed cutter of plough ahead of mouldboard (sock). Similarly used on
    other tools
Shock   alternative name for stook
Sickle   curved knife usually for reaping
Silage   grass stored green for stock feeding
Singletrees   *see* whippletrees

Skim coulter   small cutter turning over top of slice ahead of plough share
Slade   part of base of plough
Slasher   long handled billhook
Smock mill   a mill of mostly wood construction, on which only the cap turns
Sock   *see* share
Spark ignition engine   internal-combustion engine in which an electric spark
   ignites the fuel mixture in the cylinder
Spit   mechanical arrangement to turn meat in front of fire
Spuds   slats across wheels to provide grip
Steam ploughing   usually with cable-drawn plough, but also steam traction
Steam tractor   steam engine towing implement
Steerage hoe   multiple horse-drawn hoe with sideways adjustment
Stilt   plough handle (neck, tail)
Stook   cut and tied corn, stacked in pyramid form in field
Stradsticks   *see* whippletrees
Straw walker   reciprocating elevator in combine harvester
Strickle   *see* riff
Strip farming   medieval method, where peasants had strips of land in shared
   fields
Stump-jump plough   *see* plough
Subsoiler   deep action plough-type tool
Swath (swathe)   cut hay spread to dry
Swingle   the beater of a flail
Swingletrees   *see* whippletrees

Tail corn   small grains
Tailings   wasted grain after threshing
Tail race   water below a water wheel
Ted   to scatter hay to allow it to dry
Tedder   a machine for tedding
Tedding   tossing and spreading hay to encourage drying
Three-point linkage   method of mounting implement on tractor
Threshing   beating grain from ears of corn
Throw hook   *see* hay bond twister
Tide mill   a water mill using the rise and fall of the tide
Tine   projecting spike
Topping   cutting green crown off sugar beet or similar crop
Tower mill   a wind mill built of brick or stone, on which only the cap turns
Tractor   four-wheeled general-purpose internal-combustion-engined alter-
   native to horse
Tractor, Caterpiller   crawler tractor by the firm of that name only
   crawler   tracked version, with endless belts around wheels
   steam   self-propelled steam engine to pull implements
Trug   wooden basket
Tumbril   cart

Turn-furrow   *see* mouldboard
Turnover plough   *see* plough
Turnwrest   *see* plough

Undershot wheel   water mill with water led to the bottom of the wheel

Wad   alternative name for cock
Waggon (wagon)   four-wheeled horse-drawn vehicle
Wain   alternative name for waggon
Ware potatoes   the largest when sorted
Water mill   *see* overshot, breast, undershot, tide
Wattle hurdle   hurdle made by interweaving hazel and similar strips
Wheelwright   builder of waggon and carts
Whin   Scottish name for gorse
Whippletrees   bars to keep draught chains apart (whiffletrees, swingletrees, singletrees, stradsticks, badikins)
Wimble   *see* hay bond twister
Windmill   *see* post mill, smock mill, tower mill
Windrow   cut grass spread to dry
Wing   *see* mouldboard

Bibliography

With agriculture as the main occupation of the majority of people throughout history until the Industrial Revolution, it is understandable that farming in all its aspects has attracted the attention of many authors, who have either described the scene as they saw it or have used their writings to expound ideas of their own. Consequently there has been a considerable amount of published material, and the development of agriculture since the invention of printing has been fairly well documented. Of course, many of the earlier books are not generally available, but it may be possible to consult them in libraries. Following is a short list of relevant books known to the author, that he thinks may be of interest to readers of this book who want to know more of the subject.

Eighteenth Century and earlier
Blith, W. *English Improver Improved* (1653)
Ellis, Wm. *The Farmers Instructor* (2nd ed 1750)
Tull, J. *New Horse Houghing Husbandry* (1733)

Nineteenth Century
  Facts of Observations Relative to Sheep, Wool etc (1809)
  General View of the Agriculture of . . . several county titles (1808)
Copland, S. *Agriculture Ancient and Modern* (1866)
Donaldson, J. *British Agriculture* (1860)
Finlayson, J. *British Farmer* (1825)
Loudon, J. C. *Encyclopaedia of Agriculture* (1831)
Morton, J. C. *Cyclopaedia of Agriculture*, 7 vols (1856)
Ransome, J. A. *Implements of Agriculture* (1843)
Wilson, Rev J. N. *Rural Cyclopaedia* (1849)

Twentieth Century
  Young Farmers' Club Booklets (series of 24)
Blandford, P. W. *Country Craft Tools* (Newton Abbot, 1974)
Cripps, A. *The Countryman Rescuing the Past* (Newton Abbot, 1973)
Fraser, C. *Harry Ferguson* (1972)

Fussel, G. E. *The Farmer's Tools, AD 1500–1900* (1952)
Partridge, M. *Farm Tools Through the Ages* (1973)
Passmore, J. B. *The English Plough* (Oxford, 1930)
Shippen & Turner. *Basic Farm Machinery* (1966, rev 1973)
West, L. A. *Agriculture: Hand Tools to Mechanization* (HMSO, 1967)
White, K. D. *Agricultural Implements of the Roman World* (Cambridge, 1967)
Whitlock, R. *Farming From the Road* (1967)
Winter, G. *A Country Camera* (Newton Abbot, 1966, repr 1972)
Wright, P. *Old Farm Implements* (Newton Abbot, 1961, repr 1974)
—— *Old Farm Tractors* (Newton Abbot, 1962, repr 1974)

Acknowledgements

The author is glad to be able to record his appreciation of the help given by the following firms in the supply of photographs: British Leyland, plates 7 and 28; Ransomes, Sims & Jefferies, plates 3, 12 and 16.

All other photographs, and all drawings, are by the author.

Index

Agricultural engines, 32
Albone, Don, 32
All-iron plough, first, 54
American tractors, 33
Animal power, 16
Animal-powered machines, 18
Arbuthnot, 56
Axes, 150

Bagging hook, 115
Balance plough, 61, 68
Baler, 140; pick-up, 141
Barley hummeller, 159
Beet, 81, 146
Bell, Rev Patrick, 119
Bells, 156
Billhooks, 96
Bird scarers, 82
Blith, Walter, 52, 64
Branding irons, 157
Breast plough, 90
Broadcast sowing, 74
Buhrstone mills, 162
Butter churn, 171
Buttermaking, 171
Buttermilk, 172
Butterworker, 172

Cable ploughing, 27, 39, 67
Cambridge roller, 98
Carts, 19
Cas chrom, 43

Cattle bells, 156
Cereal harvesting, 113
Chaff cutter, 159
Chain-link harrows, 105
Chain saw, 152
Chain traces, 50
Cheese: making, 173; press, 174; taster, 174; vat, 173
Chessart, 173
Churn: butter, 171; milk, 170
Cider making, 165
Cock, hay, 139
Combine harvesters, 131
Compression-ignition engines, 31
Cooler, milk, 170
Corn crusher, 161
Coulter, 47; disc, 49; skim, 49, 56
Covings, 127
Cradle, scythe, 116
Cream, separating, 171
Crook, shepherd's, 155
Crop spraying, 83
Crusher, corn, 161
Cultivating, 86
Cultivation, machine, 98
Cultivator, 107
Curds, 173
Cutter, root, 161
Cutting grass, 135

Dairying, 168
Dairy produce, 166

185

Deere, John, 60
Dibber, 74
Diesel, 31; engines, 31
Digging: mechanical, 108; stick, 43
Disc coulter, 49
Disc harrow, 105
Disc mower, 136
Disc plough, 66
Ditches, 148
Dog, log, 153
Draining plough, 64
Drill, seed, 78
Drying hay, 137
Dung, 83

Early agricultural engines, 32
Earth breaking, 90
Elevator, 139
Estate management, 148

Feeding, stock, 155
Fences, 148; wire, 148
Ferguson, Harry, 40
Fertilizing, 83
Fiddle, 74
Flail, 126
Food preparation, 157
Foot plough, 43
Forage harvester, 142
Ford, Henry, 35
Forestry, 149
Forks, 90
Fowler, John, 65, 67
Froe, 152

Grass: cutting, 135; gathering, 136
Guernsey plough, 54

Hake, 50
Hammer mill, 162
Hand gathering tools, 122
Harrows, 102; chain-link, 105; disc, 105
Harvester, forage, 142
Harvesting: cereals, 113; potatoes, 142; sugar beet, 146

Hay bond twister, 124
Hay cock, 139
Hay drying, 137
Haymaking, 134
Hay sweep, 138
Hedger, 149
Hedges, 148
Hedge trimmers, 148
Hedging, 148
Hertfordshire plough, 53
Hoe, rotary, 108
Hoe, steerage, 111
Hoes, 94, 110
Hole boring, 96
Hook, sack, 164
Horse power, 29
Horses, 16
Hummeller, 159
Hussey reaper, 120

Industrial Revolution, 13
Internal-combustion engines, 30
Irons, branding, 157
Ivel tractors, 32

Jefferson, Thomas, 56

Kent plough, 52
Kibbling, 161

Laval, Dr Gustave De, 171
Levelling box, 102
Log dog, 153

Machete, 96
Machine cultivation, 98
Management: estate, 148; stock, 155
Making butter, 171
Manning, William, 135
Manure, 83
Massey-Harris, 41
Mattock, 91
McCormick, Cyrus, 120
Measuring, 164
Mechanised cutting, 117
Milk churn, 170

Milking, 166
Milking machines, 166
Milk pail, 170
Milk separator, 171
Mill: hammer, 162; oilcake, 162; plate, 162
Modern tractors, 41
Mogul tractor, 37
Mole plough, 26, 64
Mouldbaert, 100
Mouldboard, 48
Mower, disc, 136
Mowing, 134
Mowing machines, 135
Muck spreading, 84
Multi-furrow plough, 72

Nib, 153
Norfolk plough, 53
Nourse, Joel, 60

Oilcake mill, 162
Oil engines, 30
One-way plough, 53
Otto cycle, 31

Pail, milk, 170
Paring plough, 64
Parbuckle, 154
Petrol engines, 30
Pitchfork, 124
Pick-up baler, 141
Planting, 74; potatoes, 79
Plate mill, 162
Plough, balance, 61, 68; breast, 90; digging, 57; disc, 66; draining, 64; foot, 43; Guernsey, 54; hand, 44; Hertfordshire, 52; Kent, 52; long-plate, 58; mole, 26, 64; multi-furrow, 63, 72; one-way, 53, 61; paring, 64; parts, 44; reversing, 63, 73; ridging, 66; riding, 67; Rotherham, 55; Sussex foot, 53; tractor, 70; turnover, 63; turnwrest, 53; Western, 50
Ploughing, 43; cable, 27; engines, 27;

steam, 27, 67
Pneumatic tyres, first, 36
Post-hole borer, 97
Potato: cleaning, 144; harvesting, 142; planting, 79; spinner, 142
Power take-off (PTO), 27, 36
Prairie breaker, 66
Preparation of food, 157
Produce, dairy, 166

Rakes, 93, 122; dump, 138; horse, 138
Reaping hook, 115
Reaping machines, 117; self-binding, 124
Reversing ploughs, 73
Ridged roller, 98
Ridging plough, 66
Riding plough, 67
Riff, 116
Roman harvesting, 117
Rollers, 98; Cambridge, 98; ring, 98
Root cutter, 161
Rotherham plough, 55
Rural scene, 11

Sack hook, 164
Sail reaper, 122
Saws, 152
Scotch hands, 172
Scythe, 115; cradle, 115
Seamer, 50
Seed drill, 78
Seed fiddle, 74
Seed-lip, 74
Self-binding reaping machines, 124
Separating cream, 171
Share, plough, 47
Shears, 155
Sheep bells, 156
Shepherd, 155
Shepherd's crook, 155
Shovels, 90
Sickle, 113
Silage, 140
Skim coulter, 49, 56
Slasher, 96

Sowing, 74
Spades, 86
Spinner, potato, 142
Spraying, 83
Steam: engines, 24; ploughing, 27; power, 16, 24; tractors, 28
Steerage hoe, 111
Stilts, plough, 49
Stock feeding, 155
Stock management, 155
Straw walker, 131, 139
Strike, 164
Sugar beet, 81; harvesting, 146
Swath, 134; turner, 137
Sweep, hay, 138
Swing plough, 46

Tedder, 136
Tedding, 136
Three-point linkage, tractor, 40, 71
Threshing, 126; machine, 127
Titan tractor, 37
Traces, chain, 50
Tractors, 30; American, 33; Ivel, 32; modern, 41; post-World War I, 34; steam, 28
Tractor, three-point linkage, 40, 71; plough, 70; walking, 40

Trevithick, Richard, 24
Tull, Jethro, 52, 77
Turnip knives, 157
Turnover plough, 63
Turnwrest plough, 53

USA pioneer farming, 60

Vallus, 117
Vat, cheese, 173
Veterinary work, 157

Waggon, 19, 22
Walking tractor, 40
Watt, James, 24
Weeding hook, 93
Western ploughs, 45, 50
Wheeled transport, 19
Wheel parts, 20
Wheelwright, 20
Whey, 173
Whin bruiser, 159
Whippletrees, 50
Wimble, 124
Winnowing, 129
Working horses, 16

Yoke, 170